Frontiers of Digital Transformation

Kazuya Takeda · Ichiro Ide · Victor Muhandiki
Editors

Frontiers of Digital Transformation

Applications of the Real-World Data Circulation Paradigm

Editors
Kazuya Takeda
Institutes of Innovation for Future Society
Nagoya University
Nagoya, Aichi, Japan

Ichiro Ide
Mathematical and Data Science Center
Nagoya University
Nagoya, Japan

Victor Muhandiki
Institutes of Innovation for Future Society
Nagoya University
Nagoya, Japan

ISBN 978-981-15-1360-2 ISBN 978-981-15-1358-9 (eBook)
https://doi.org/10.1007/978-981-15-1358-9

This Springer imprint is published by the registered company Springer Nature Singapore Pte Ltd.
The registered company address is: 152 Beach Road, #21-01/04 Gateway East, Singapore 189721, Singapore

Preface

The speed of data transmission between computers surpassed that of human communications long ago, and has since expanded exponentially. As a result, the origin of the majority of data has become non-human, mechanical, or natural sources; in fact, humans are merely the source of a small part of the current data explosion. Such expanding data transmission does not simply consist of single source and destination pairs, but actually circulates over a complex network connecting numerous sources and destinations. We should note that such circulation is an important aspect of the underlying systems. For example, in engineering, it is well known that a feedback loop can stabilize a dynamic system. This fact implicates the possibility of controlling the torrential flow of data circulation by human intervention even on a small amount of data. Based on this concept, in order to tame and control the massive amount of data originating from non-human sources, we have been considering the insertion of "acquisition," "analysis," and "implementation" processes in the flow of data circulation.

Although this approach has the potential to provide many societal benefits, data circulation has not typically been the target of academic research. Thus, in 2013, we started a new degree program in this domain, namely, Real-World Data Circulation (RWDC), gathering faculty and students from the Graduate Schools of Information Science, Engineering, Medicine, and Economics in Nagoya University, Japan.

This book is the first volume of a series of publications summarizing the outcome of the RWDC degree program, collecting the relevant chapters of graduate students' dissertations from various research fields targeting various applications, as well as lecture notes in relevant fields. Throughout the book, we present examples of real-world data circulation and then illustrate the resulting creation of social value.

Nagoya, Japan
January 2021

Kazuya Takeda
Ichiro Ide
Victor Muhandiki

Contents

Contributors

Bao Trung Chu Super Zero Lab, Tokyo, Japan

Hatem Darweesh Nagoya University, Nagoya, Japan

Mari Endo Kinjo Gakuin University, Nagoya, Japan

Tomoki Hayashi Human Dataware Lab, Nagoya, Japan

Sheng Hu Hokkaido University, Sapporo, Japan

Kohei Isechi Nagoya University, Nagoya, Japan

Hongjin Jung Korean Institute of Machinery & Materials, Daejeon, South Korea

Shimeng Peng Nagoya University, Nagoya, Japan

Shogo Seki Nagoya University, Nagoya, Japan

Kazuya Takeda Nagoya University, Nagoya, Japan

Takahiro Tsukamoto Nagoya University, Nagoya, Japan

Chenxi Tu Huawei Technologies, Shenzhen, China

Takahiro Yamakoshi Nagoya University, Nagoya, Japan

Xinxian Zhang Beihang University, Beijing, China

Introduction

Introduction to the Real-World Data Circulation Paradigm

Kazuya Takeda

1 Real-World Data Circulation (RWDC)

The essential social value is formed by widely sharing the fundamental values such as convenience, enjoyability, well-being, and affluence with other people. Such values are not simply delivered from the product/service creators to the consumers but are created through the interactive processes of both creators' ideas and consumers' demands. The consumers' demands are usually not visible and grow/change during the use of new products and services. Circulations that connect users and creators that enable creating new products/services that reflect the ever-changing or unconstructed users' demands well are indeed the essential social value creation processes. The lack of attention to this circulation may be one of the reasons that has caused Japan's degradation in the global competitiveness ranking[1] from the 1st position (1990) to the 24th (2013), and 34th (2020).

We believe, by the following two reasons, that in order to create such a circulation, a new research paradigm is needed: (1) creating new social values essentially implicates a multi-disciplinary study involving at least engineering (convenience), computer science (enjoyment), medicine (health), and economics (abundance), and (2) connecting creators and consumers inevitably requires three steps such as sensing the demands from the measurement of the real world (data acquisition), analyzing the data to understand demands (data analysis), and based on the hypothesis derived from the understanding, modifying or even newly creating products/services (implementation). Needless to say that the implantation of the new products/services may affect the customers' behaviors which would be measured again by the first step (data

[1] International Institute for Management Development, "World competitiveness ranking," https://worldcompetitiveness.imd.org/ [Accessed: Jan. 11, 2021].

K. Takeda (✉)
Nagoya University, 1 Furo-cho, Chikusa-ku, Nagoya 464-8601, Japan
e-mail: kazuya.takeda@nagoya-u.jp

K. Takeda et al. (eds.), *Frontiers of Digital Transformation*,
https://doi.org/10.1007/978-981-15-1358-9_1

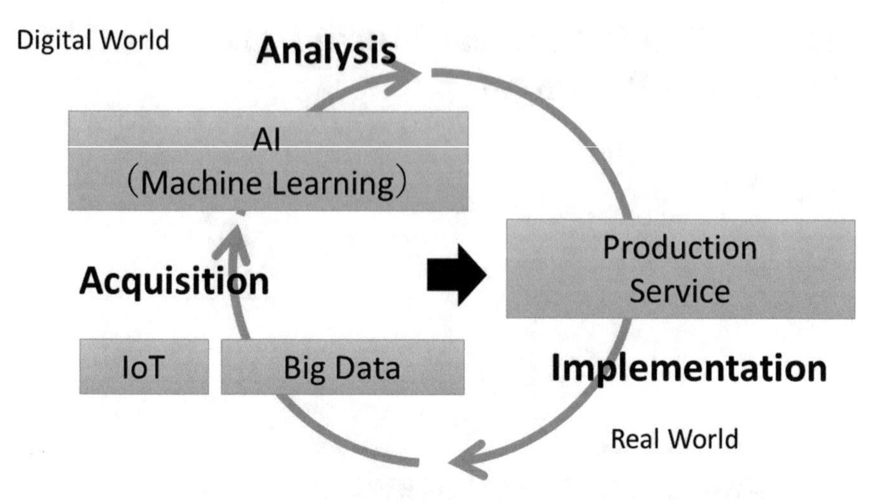

Fig. 1 The RWDC process is indispensable in order to transform the production process and/or consumer services using digital technologies

acquisition). The new academic discipline, Real-World Data Circulation (RWDC), is a discipline that studies this circulation imposed in either academic or industrial applications. (Fig. 1)

Fostering Ph.D. level experts of RWDC is crucially important for the Japanese industry where less attention has been paid to the system technologies rather than the production technologies. For example, it is well known that although many parts inside Apple's iPhone are "Made in Japan," very few apps are provided by Japanese companies. The industrial standing point is clearly contrasting to that of the US and China.

Here let me show one example of the RWDC skillset. This is a job description of a growth engineer (In general, a growth hacker) given by Dropbox almost a decade ago. Dropbox at that time was an emerging startup company and therefore their human demand was somehow projecting the current social needs of human resource.

> A growth engineer would substantially contribute to Dropbox's continued success. The process is simple: measure everything to understand it, come up with new ideas, test the best ones, launch the best performing, and repeat this all as quickly as possible.

Inspired by this job description as well as the common understanding of the importance of RWDC, four graduate schools of Nagoya University, Japan jointly put a unique proposal to the call for "Program for Leading Graduate Schools" by the Ministry of Education Culture, Sports, Science and Technology (MEXT), in 2012.

This book introduces essences of the RWDC considered in Ph.D. theses of 13 students who joined the program from various research fields and disciplines during 2014 and 2021 in order to showcase the actual cases of RWDC.

2 Real-World Data Circulation in the Human, Machine, and Society Domains

This section will overview the dissemination of the RWDC concept in each of the human, machine, and social domains that are showcased in this book.

2.1 RWDC in the Human Data Domain

As examples of RWDC in the human data domain, four research topics are showcased in Part II of this book. First, two topics on auditory scene analysis titled "A study on environmental sound modeling based on deep learning" and "A study on utilization of prior knowledge in underdetermined source separation and its application" are introduced. Next, a topic on computer-aided education titled "A study on recognition of students' multiple mental states during discussion using multimodal data" is introduced. In the end, a topic on information security titled "Towards practically applicable quantitative information flow analysis" is introduced.

2.2 RWDC in the Machine Data Domain

As examples of RWDC in the machine data domain, five research topics are showcased in Part III of this book. First, a topic on material processing titled "Research on high-performance high-precision elliptical vibration cutting" is introduced. Next, two topics on coding of optical sensor data titled "A study on efficient light field coding" and "Point cloud compression for 3D LiDAR sensor" are introduced. Thirdly, a topic on route planning for autonomous driving titled "Integrated planner for autonomous driving in urban environments including driving intention estimation" is introduced. Finally, a topic on fluid dynamics analysis titled "Direct numerical simulation on turbulent/non-turbulent interface in compressible turbulent boundary layers" is introduced.

2.3 RWDC in the Social Data Domain

As examples of RWDC in the social data domain, four research topics are showcased in Part IV of this book. First, two topics on correction of real-world text data titled "Efficient text autocompletion for online services" and "Coordination analysis and term correction for statutory sentences using machine learning" are introduced. Next, a topic on the analysis of cityscape analysis titled "Research of ICT utilization for consideration of townscapes" is introduced. Finally, a topic on industrial science titled "Measuring efficiency and productivity of Japanese manufacturing industry considering spatial interdependence of production activities" is introduced.

3 Human-Resource Development in the RWDC Domain

This section introduces the four pillars of human-resource development in the RWDC domain as a degree program.

3.1 Ph.D. Research

The human resource development in the RWDC Ph.D. program consists of four pillars (Fig. 2). The first pillar, of course, is the Ph.D. research experiences in one of the four graduate schools. The second pillar is the knowledge and skills in RWDC, which will be discussed in detail in Sec. 3.2. The third pillar is the global experiences such as student collaboration, visiting research, and networking. Finally, the fourth pillar is the industrial experiences including startup experiences. For general students, pursuing all four requirements in a 5-year period is not easy, since the latter three pillars should be fulfilled on top of the first pillar, i.e., Ph.D. research. As most Ph.D. programs in Japanese graduate schools, it is a mandatory requirement for our students

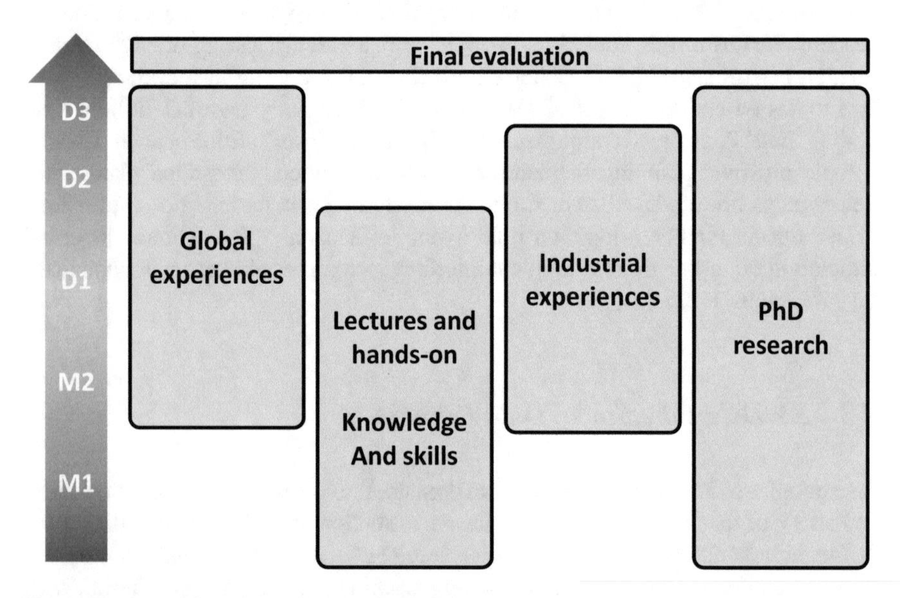

Fig. 2 The four pillars of the education in the RWDC program

to publish multiple journal papers. In other words, the RWDC Ph.D. program requires extra achievements to the Ph.D. degree.

In the RWDC Ph.D. program, acquisition, analysis, and implementation are not independent procedures but is one general process as a whole. Whatever the topic of his/her Ph.D. thesis may be, the new findings should be connected to an RWDC and therefore have to have impacts to our society. In order to guarantee the value of the degree of the program, a degree committee is formed and requests the applicant student to add an additional chapter to the Ph.D. thesis where he/she discusses the research achievement in terms of RWDC. Some experts from the industry are also invited for reviewing that chapter.

3.2 Knowledge and skills in RWDC

Dealing with real-world data has increased its importance in the past 20 years, and will definitely continue to increase toward the future. We believe that our RWDC Ph.D. program was one of the pioneering attempts emphasizing the importance of *real-world* data circulation. In general, the RWDC consists of three steps: the first step is data acquisition, i.e., acquiring data from the real world. This process can be generalized as the process of "real to virtual information transformation" in a wider view. There are various means of data acquisition such as physical sensing with IoT devices, SNS statistics, public open data, private management indicators, etc. Students are requested to study the fundamental theories and skills of data acquisition in two of their majoring target domains through recommended lectures taught in the four member graduate schools.

In the RWDC program, we categorize the target of RWDC into three domains: human, machine, and society. Students are requested to select two out of the three categories as primary and secondary domains. According to their target domains, students can put emphases in some particular acquisition methods.

The second process is data analysis. Students start with learning fundamental theories of statistics and signal and pattern information processing as well as machine learning. Then, they study various cases of data analysis applications through lectures taught by professors in the four member graduate schools. This lecture series, namely, Real-World Data Circulation Systems I, connects theories/methods of data analysis applied to various real problems in research/technology.

The third process of the RWDC is implementation. In order to interact with, we finally implement the analysis results to the real world, which initiates the next circulation. That part is taught mainly by invited lecturers from the industry.

3.3 *Global/Industrial Leadership*

Unlike other Ph.D. programs, the RWDC Ph.D. program puts emphasis on the skill development, particularly for becoming a global leader. It is not easy to enhance such skills based solely on lessons such as practicums, group works, and discussions held in classrooms. The designed education style is, so-to-say "Giving the students opportunities and evaluate their experiences." Thus, the program mandates all students to experience research activities in global environments during a minimum of 2-month long visiting research (some students would actually stay longer, such as 1 year) outside their home country. In addition to that, students must participate in a 2-week long summer school held in Asian countries during the second year in the program. The summer schools have been located in Istanbul, Turkey (2015); Hanoi, Vietnam (2016 and 2018); and Bangkok, Thailand (2017 and 2019). Collaborating with the students of Istanbul Technological University (ITU), Hanoi Institute of Science and Technology (HIST), and Chulalongkorn University, our students designed cultural lectures with hands-on activities, together with short project involving group works. We found that these global experiences brought a big change to our students in the sense that it eliminates the fears for communicating and collaborating with people from different cultures and backgrounds in English. In fact, it is not necessarily difficult, but they have not been aware of how easy it is until they actually tried out. Of course, staying in the World's top-level laboratories even for a short period of time connected our students with the premier research community which would become an eternal asset for the young researchers.

3.4 *Industrial Experience*

The program also requests all students to take part in industrial internships for at least 2 months. The experiences in time and goal managements in a corporate project sometimes changed the students' attitude drastically.

4 Achievements of the RWDC program

The research achievements obtained through the RWDC program will be evident in the following chapters. We will see various data circulations formed upon the Ph.D. researches in different disciplines, spanning from material science to social economics. Each of them has an associated circulation(s), and therefore tightly connected to the real world. These chapters will showcase the effectiveness of our systematic scheme in education that combines practical experiences and deep research. This is obviously the most important achievement of the program. In the future, the pile of research discussions in the following chapters will serve as the foundation of a new research discipline, Real-World Data Circulation, fostering young leaders in that area.

Another symbolic achievement throughout the first 7 years was the young talents themselves. For instance, we are proud of the fact that ten startups were launched from the program. As of summer 2020, they have raised more than ten million USD of funds and have created more than 100 job opportunities. Many students in the program are closely connected with those startups, and they experienced that research achievements can work as parts of the industrial eco-system once they are used in the context of RWDC. The importance and the social demand of our targeted talent, RWDC leaders, are still growing. Actually, it is attracting more attention now than the time we designed the program as we can see from the fact that data scientists still seem to be one of the most popular job titles among employers; talents who can collect data, analyze them, and recommend tactics; who understand statistics; who knows which analysis model works on which problem; and who can find the appropriate computer tools or write codes for that. On top of that, with the understanding of the crucial importance of RWDC, as well as the deep knowledge on a specific discipline, our students can surely pioneer the next step needed for the innovation toward our future society.

5 Future Prospective of the Program—Beyond Digital Transformation

The concept of RWDC may not sound very new to some people. They may wonder how it differs from the concept of Plan-Do-Check-Act (PDCA). To me, RWDC is different from PDCA in its final goal. PDCA is a self-governing process toward a given goal, while RWDC is an ever-continuing innovation process. PDCA or feedback control is a nice idea if we wished to stabilize complex systems, while RWDC is a fundamental policy of trying to do new things where we wish to change as much as possible.

Recently, the concept of Digital Transformation (DX) is becoming popular in Japan. Sometimes it simply indicates the application of digital technologies such as IoT, Big data, and AI for improving the efficiency of an organization, particularly a company's production or service systems. But to me, it is obvious that the biggest advantage of using digital technologies is enlarging and accelerating the deformation process, which is not in the form of a single pipeline, but rather is a circulation or an interactive process between the technologies and the human society. We are proud that RWDC has deeply recognized the importance of the fact that transformation is the result of interaction, and that we have continuously worked in the education of the leaders who share this concept.

Although in this volume, the number of chapters, i.e., Ph.D. theses produced through the RWDC program, is limited to 13, our continuous efforts will extend the chapters in succeeding volumes published in the future, which will keep on contributing to build the new discipline of RWDC, as well as producing global leaders in industrial science.

Frontiers in Human Data Domain

A Study on Environmental Sound Modeling Based on Deep Learning

Tomoki Hayashi

Abstract Recent improvements in machine learning techniques have opened new opportunities to analyze every possible sound in the real-world situation, namely, understanding environmental sound. This is a challenging problem because the goal is to understand every possible sound in a given environment, from the sound of glass breaking to the crying of children. This chapter focuses on Sound Event Detection (SED), one of the most important tasks in the field of understanding environmental sound, and addresses three problems that affect the performance of monophonic, polyphonic, and anomalous SED. The first problem is how to combine multi-modal signals to extend the range of detectable sound events into human activities. The second one is how to model the duration of sound events that is one of the essential characteristics to improve polyphonic SED performance. The third one is how to model normal environments in the time domain to improve anomalous SED systems. This chapter introduces how the proposed method solves each problem and reveals the effectiveness of the proposed method to improve the performance of each SED task. Furthermore, discussions about the relationship between each work and the Real-World Data Circulation (RWDC) reveal how each work accomplishes what kind of data circulation.

1 Introduction

Humans encounter many kinds of sounds in daily life, such as speech, music, the singing of birds, and keyboard typing. Over the past several decades, the main targets of acoustic research have been speech and music, while other sounds have generally been treated as background noise. However, recent improvements in machine learning techniques have opened new opportunities to analyze such sounds in detail, namely, understanding environmental sound. Understanding environmental sound is challenging because the goal is to understand every possible sound in a given environment, from the sound of glass breaking to the crying of children. To accelerate

T. Hayashi (✉)
Human Dataware Lab. Co., Ltd, Nagoya, Japan
e-mail: hayashi@hdwlab.co.jp

K. Takeda et al. (eds.), *Frontiers of Digital Transformation*,
https://doi.org/10.1007/978-981-15-1358-9_2

this research field, several competitions have been held in recent years, including CLEAR AED [41], TRECVID MED [34], and the DCASE Challenge [31]. Moreover, several datasets have been developed, such as urban sound datasets [39] and AudioSet [13].

One of the essential tasks in this field is Sound Event Detection (SED), which is the task to detect the beginning and the end of sound events and to identify their labels. SED has many applications such as retrieval from multimedia databases [48], life-logging [36, 42], automatic control of devices in smart homes [45], and audio-based surveillance systems [6, 7, 44]. The SED task can be divided into three types: monophonic, polyphonic, and anomalous SED. An overview of each SED task is shown in Fig. 1. An acoustic feature vector (e.g., log Mel filterbanks or Mel-frequency cepstral coefficients) is extracted from the input audio clip, the length of which is around several minutes. Then, the SED model estimates the beginning and the end of sound events and identifies their labels from a given feature vector. The target sound events are predefined, but it depends on the SED type. In the monophonic SED case, multiple sound events cannot appear at the same time. On the other hand, in the polyphonic SED case, any number of sound events can be overlapped at the same time. In the anomalous SED case, the prior information of target sound events is not given. Therefore, the system tries to detect novel or anomalous sound events which do not appear in the training data. Sound events include a wide range of phenomena that vary widely in acoustic characteristics, duration, and volume, such as the sound of glass breaking, typing on a keyboard, knocking on doors, and human speech. This diversity of targets makes SED challenging. Though recent advances in machine learning techniques have led to the improvement of the performance of SED systems, various problems remain to be solved.

This study addresses three problems that affect the performance of monophonic, polyphonic, and anomalous SED systems. The first problem is how to combine multi-modal signals to extend the range of detectable sound events into human activities. The second is how to model the duration of sound events that is one of the essential characteristics to improve polyphonic SED performance. The third is how to model normal environments in the time domain to improve anomalous SED systems.

First, toward the development of a life-logging system, the use of multi-modal signals recorded under realistic conditions is focused [16, 19]. In that way, sound events related to typical human activities can be detected, including discrete sound events like a door closing along with sounds related to more extended human activities like cooking. The key to realizing the application is finding associations between different types of signals that facilitate the detection of various human activities. To address this issue, a large database of human activities recorded under realistic conditions is created. The database consists of over 1,400 h of data, including the outdoor and indoor activities of 19 subjects under practical conditions. Two Deep Neural Network (DNN)-based fusion methods using multi-modal signals are proposed to detect various human activities. Furthermore, the speaker adaptation techniques in Automatic Speech Recognition (ASR) [32] are introduced to address the subject individuality problem, which degrades the detection performance. Experimental results using the constructed database demonstrate that the use of multi-modal signals is effective,

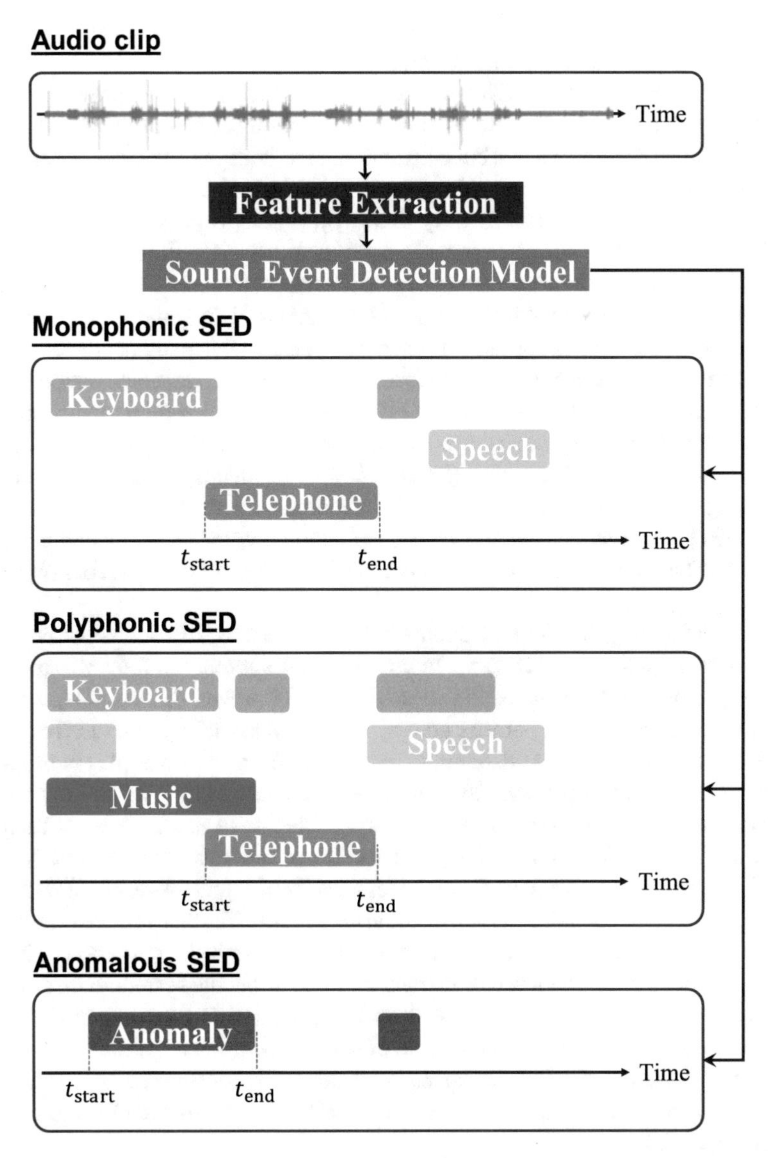

Fig. 1 Overview of the sound event detection task. First, the system extracts a feature vector from the input audio clip. Then, the SED model estimates the beginning and the end of sound events and their labels from the feature vector. Here, the target types of sound events are predefined. In the monophonic case, a sound event cannot be overlapped. Meanwhile, in the polyphonic case, any number of sound events can be overlapped. In the anomalous case, the prior information on the target sound events is not given; therefore, the system detects novel or anomalous sound events which do not appear in the training data

and that speaker adaptation techniques can improve performance, especially when using only a limited amount of training data.

Second, modeling the duration of sound events is focused to improve the performance of polyphonic SED systems [17]. The duration is one of the most important characteristics of sound events, but conventional methods have not yet modeled them explicitly [21, 24]. To address this issue, a novel hybrid approach using duration-controlled Long Short-Term Memory (LSTM) [14, 22] is proposed. The proposed model consists of two components: a Bidirectional LSTM recurrent neural network (BLSTM), which performs frame-by-frame detection, and a Hidden Markov Model (HMM) or a Hidden Semi-Markov Model (HSMM) [49] that models the duration of each sound events. The proposed approach makes it possible to model the duration of each sound event precisely and to perform sequence-by-sequence detection without needing thresholding. Furthermore, to effectively reduce insertion errors, the post-processing method using binary masks is also introduced. This post-processing step uses a Sound Activity Detection (SAD) network to identify segments for activity indicating any sound event. Experimental evaluation with the DCASE2016 task2 dataset [31] demonstrates that the proposed method outperforms conventional polyphonic SED methods and can effectively model sound event duration for polyphonic SED.

Third, modeling the normal acoustic environment is focused to improve the anomalous SED system [18, 25]. In conventional approaches [29, 38], the modeling is performed in the acoustic feature domain. However, this results in a lack of information about the temporal structure, like the phase of the sounds. To address this issue, a new anomalous detection method based on WaveNet [47] is proposed. WaveNet is an autoregressive convolutional neural network that directly models acoustic signals in the time domain, which enables us to model detailed temporal structures like the phase of waveform signals. The proposed method uses WaveNet as a predictor rather than a generator to detect waveform segments responsible for significant prediction errors as unknown acoustic patterns. Furthermore, i-vector [8] is utilized as an additional auxiliary feature of WaveNet to consider differences in environmental situations. The i-vector extractor should allow the system to discriminate the sound patterns, depending on the time, location, and surrounding environment. Experimental evaluation with a database of sounds recorded in public spaces shows that the proposed method outperforms conventional feature-based approaches and that time-domain modeling in conjunction with the i-vector extractor is effective for anomalous SED.

In the following sections, the relationship between each work and Real-World Data Circulation (RWDC) has been discussed. Section 2 explains how the developed tools for human activity recognition can foster RWDC. Section 3 describes how data analysis during research on polyphonic SED inspires a new method for SED based on duration modeling. Section 4 explains how the output from the research on anomalous SED can be used to improve the performance of polyphonic SED systems, demonstrating the application of discovered knowledge to another process. Finally, this chapter is summarized in Sect. 5.

2 Human Activity Recognition with Multi-modal Signals

The goal of this work is the development of a method to promote the cycle shown in Fig. 2. This cycle assumes that intellectual and physical activities result in experiences, such as the discovery of new knowledge or a sense of accomplishment. These experiences enhance people's abilities, expanding the range of activities open to them. Continuously repeating this cycle makes it possible for us to develop ourselves and improve their quality of our life. In order to help people to keep repeating this cycle, it is necessary to monitor them and to understand their activities and experiences. To achieve this, a life-logging system was developed, which automatically records the signals and recognizes human activity, from simple movements such as walking to complex tasks such as cooking. An overview of the target life-logging system is shown in Fig. 3. Users attach a smartphone that records environmental sound and acceleration signals continuously, and then the signals are sent to the server. The server receives the signals and recognizes the subject's current activity by the proposed human activity recognition model. Finally, the results are then sent to the subject's smartphone. The subjects can not only view their activity history but also send feedback to improve recognition performance. Furthermore, the system can provide a recommendation of the activity based on their history.

There are two important points to develop such applications: the usability of the system and the range of recognizable activities under realistic conditions. To address the first point, a smartphone-based recording system was developed. The smartphone-based system does not require attaching a large number of sensors, easy and less burden to use. To address the second point, a large database of human activity consisting of multi-modal signals recorded under realistic conditions is created. Then, two DNN-based fusion methods that use multi-modal signals were developed to recognize complex human activities. Furthermore, speaker adaptation techniques were introduced to address the problem of the subject individuality, which degrades the system performance when the model constructed for a particular subject is used to classify the activities of another subject. Experimental results with the constructed database demonstrated that using multi-modal signals is effective, and the speaker adaptation techniques can improve the performance, especially when using only a limited amount of training data.

Finally, a human activity visualization tool was developed by integrating the above systems, shown in Fig. 4. In Fig. 4, the right side represents the results of activity recognition, and the left side displays the recorded signals, while the center shows the monitored video and the geographic location of the smartphone user. The system can collect and analyze the individual data, and then feedback them to improve the activity recognition performance. Furthermore, it can provide recommendations to encourage users to make them more active based on the analyzed data. Thus, the developed systems enable us to promote not only the RWDC, i.e., the cycle of data acquisition, data analysis, and implementation but also the loop in Fig. 2 which improves the quality of our life.

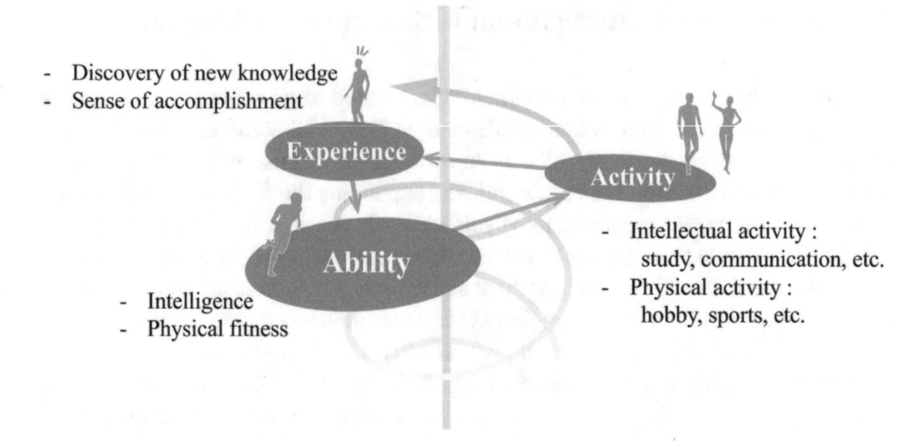

Fig. 2 Overview of the target cycle. Activities yield experiences, such as the discovery of new knowledge or a sense of accomplishment. These experiences enhance people's abilities, expanding the range of activities that can be attempted

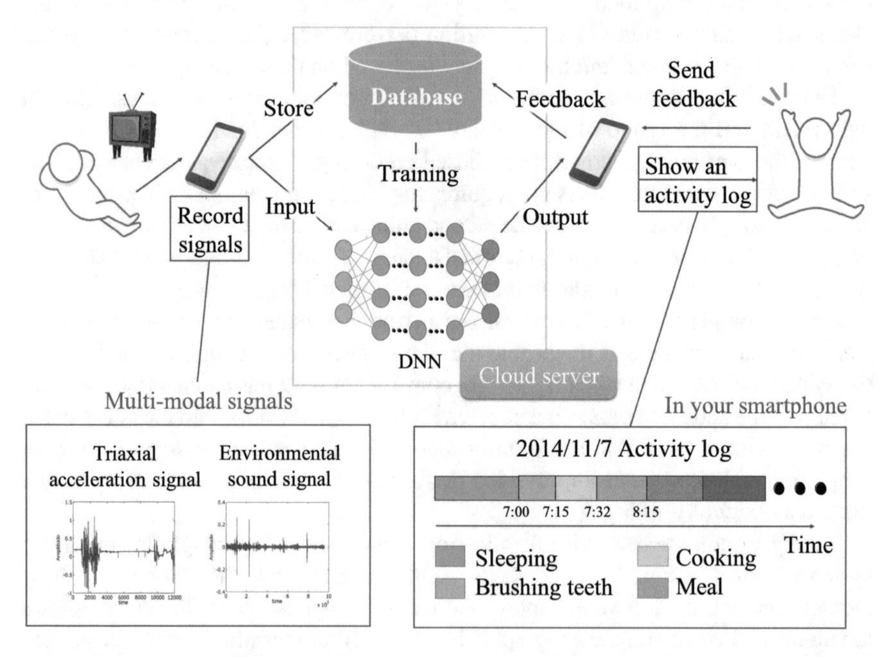

Fig. 3 Overview of the life-logging system. The system uses a smartphone to record environmental sound and acceleration signals continuously. The server receives the signals and recognizes the subject's current activity by the proposed human activity recognition model. Finally, the results are sent to the subject's smartphone. The subjects can not only view their activity history but also send feedback to improve recognition performance

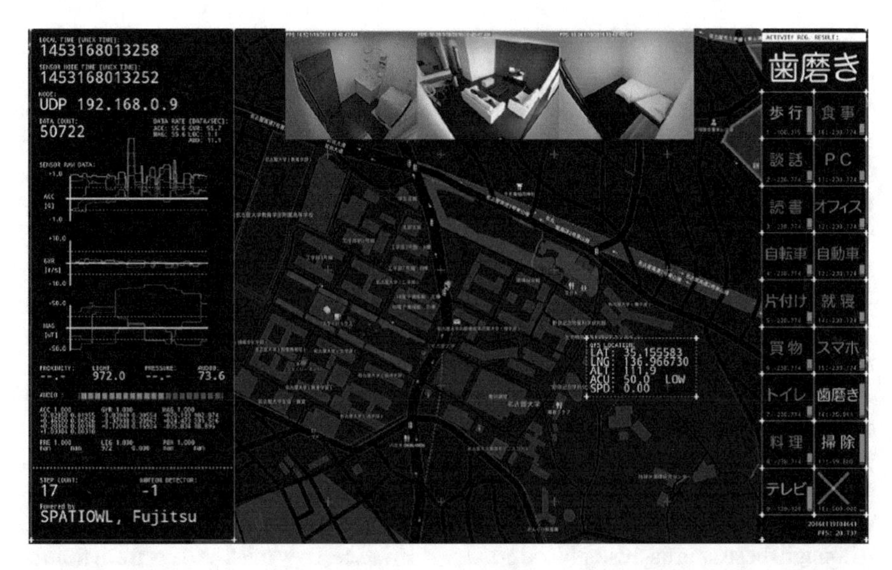

Fig. 4 Developed human activity visualization tool. On the right, the results of activity recognition are shown. On the left, the recorded signals are displayed. A monitored video is shown at the top center, and the location of the subject with the smartphone is indicated on the map in the center

3 Polyphonic SED Based on Duration Modeling

To realize practical applications, we need to analyze the characteristics of acquired data in detail and to develop a method based on these characteristics. In SED, there are various characteristics of sound events, and one of the most important characteristics is duration, which represents how the sound event continues in the time direction. The histogram of sound events is shown in Fig. 5. The horizontal axis represents the number of frames, i.e., duration, and the vertical axis represents the number of appearances. The bigger number of frames represents the duration of the sound events is longer. The figure shows that each sound event has a different duration, and it should help to improve detection performance. However, in conventional methods, this important information was not utilized explicitly.

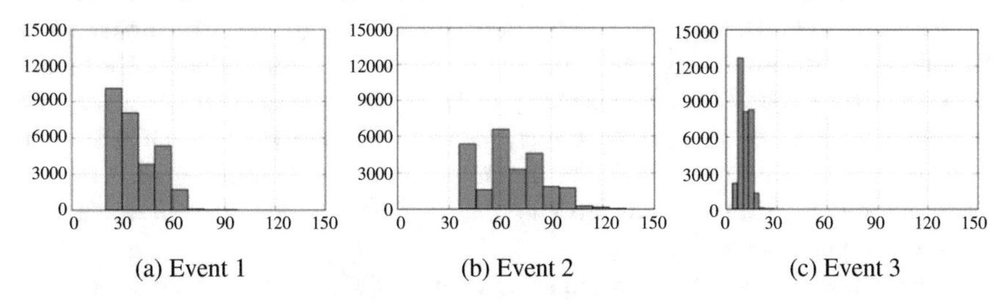

Fig. 5 Histogram of different three sound events. The horizontal axis represents the number of frames, i.e., duration. The bigger the value is, the duration of the sound event is longer

A typical conventional approach is to use Non-negative Matrix Factorization (NMF) [9, 20, 23, 24]. In NMF approaches, a dictionary of basis vectors is learned by decomposing the spectrum of each single-sound event into the product of a basis matrix and an activation matrix, and then combining the basis matrices of all the sound events. The activation matrix for testing is estimated using the combined basis vector dictionary, and is then used either for estimating sound event activations or as a feature vector that is passed on to a classifier. These NMF-based methods can achieve good performance, but they do not take correlations in the time direction into account, instead of performing frame-by-frame processing. As a result, the prediction results lack temporal stability so that extensive post-processing is needed. Moreover, the optimal number of bases for each sound event must also be identified.

More recently, methods based on neural networks have been developed, which have also achieved good SED performance [1, 3, 5, 10, 11, 35, 46]. A single network is typically trained to solve a multi-label classification problem involving polyphonic SED. Some studies [1, 11, 35, 46] have also utilized Recurrent Neural Networks (RNNs), which are able to take into account correlations in the time direction. Although these approaches achieve good performance, they must still perform frame-by-frame detection, and they do not explicitly model the duration of the output label sequence. Additionally, threshold values for the actual outputs need to be determined carefully to return the best performance. However, the conventional methods have not explicitly utilized this information.

To address this issue, a novel hybrid approach using duration-controlled LSTM was proposed, which can utilize the input sequential information and model the duration of each sound event explicitly. The proposed model consists of two components: a bidirectional LSTM that performs frame-by-frame detection and an HMM or an HSMM that models the duration of each sound event. The proposed hybrid system is inspired by the BLSTM-HMM hybrid system used in speech recognition systems [4, 15, 37]. An overview of the proposed method is shown in Fig. 6. In the proposed method, each sound event is modeled by HSMM (or HMM). The network predicts the posterior of states of each HSMM from a given acoustic feature sequence. The posterior is then converted to output probability of states using the state prior based on Bayes' theorem. Finally, Viterbi decoding is performed using the converted probability in each HSMM [49]. Thanks to the use of HSMM (or HMM), the proposed approach made it possible to model the duration of each sound event precisely and perform sequence-by-sequence detection without thresholding. Furthermore, to effectively reduce insertion errors, which often occur under noisy conditions, a post-processing step based on a binary mask was introduced. The binary mask relies on a Sound Activity Detection (SAD) network to identify segments with an arbitrary sound event activity. Experimental evaluation with the DCASE2016 task2 dataset [31] demonstrated that the proposed method outperformed conventional polyphonic SED methods, proving that modeling of sound event duration is effective for polyphonic SED.

A prediction example is shown in Fig. 7, where the top box represents the input audio signal, the middle one represents the spectrogram of the input audio signal, and the bottom one represents the SED results. In the bottom box, the purple bar

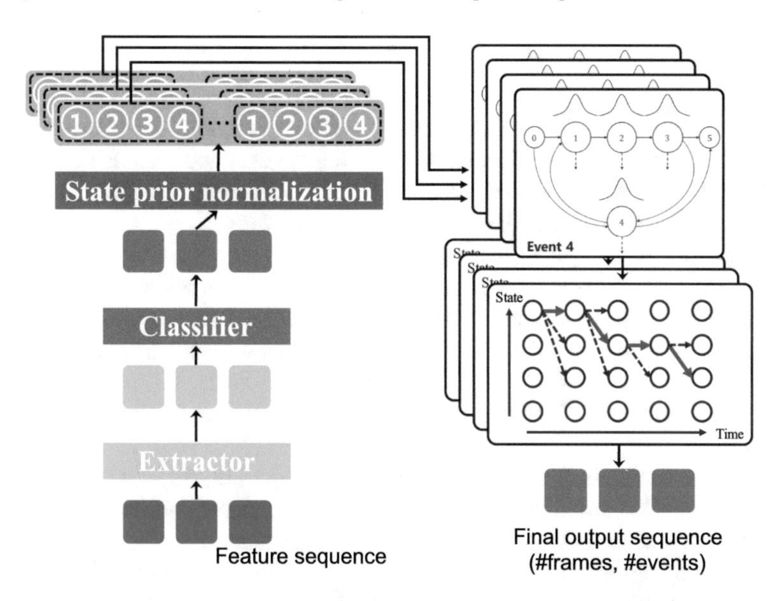

Fig. 6 Overview of the proposed duration-controlled LSTM model. Each sound event is modeled by HSMM (or HMM). The network predicts the posterior of states of each HSMM. The posterior is converted to the output probability of states, and then in each HSMM, Viterbi decoding is performed using the converted probability

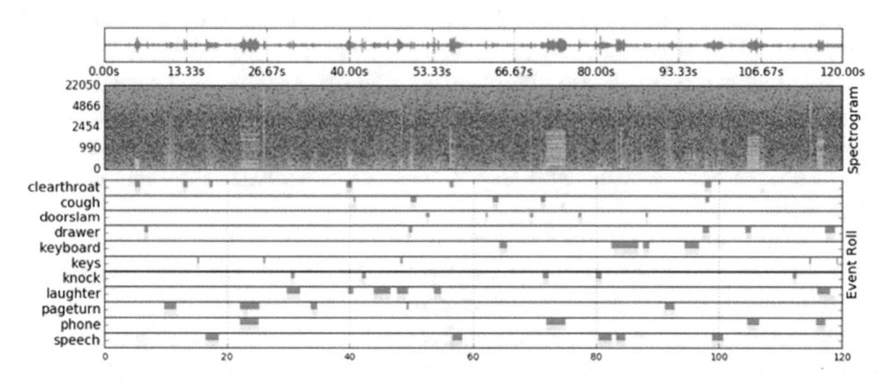

Fig. 7 Prediction example of the proposed method. The top box represents the input audio signal. The middle one represents spectrogram of the input audio signal. The bottom one represents the SED results. The purple bar represents the ground truth, while the light green bar represents the prediction results

represents ground truth, and the light green bar represents the prediction results. As we can see, the proposed model can detect sound events even if they are overlapped. Through this work, we can see how the important characteristics are analyzed and how the analyzed characteristics were integrated with the neural network. Therefore, this is a good illustration of the analysis phase of RWDC, and a new method of polyphonic SED was successfully developed through the analysis of data acquired from the real world.

4 Anomalous SED Based on Waveform Modeling

The task of anomalous SED is to identify novel or anomalous sound events from a given audio clip. The prior information about the target sound event is not provided, and therefore the anomalous SED model is trained in an unsupervised manner, which does not require the training data of the target sound events.

One of the typical unsupervised approaches is change-point detection [2, 26, 33], which compares a model of the current time with that of a previous time to calculate a dissimilarity score, and then identifies highly dissimilar comparison results as anomalies. However, in the real-world situation, the sounds which can occur are highly variable and non-stationary, and therefore the detected change points are not always related to anomalies that are of concern (e.g., the sound of the departure of the train).

Another unsupervised approach is outlier detection [12, 30, 43], which models an environment's normal sound patterns, and then detects patterns which do not correspond to the normal model identifying them as anomalies. Note that the normal patterns are patterns that have appeared in the training data. Typically, a Gaussian Mixture Model (GMM) or one-class Support Vector Machine (SVM) with acoustic features has been used [27, 40]. Thanks to recent advances in deep learning, neural network-based methods are now attracting attention [28, 29, 38]. These methods train an Auto-Encoder (AE) or an LSTM-RNN with only normal scene data. While an AE encodes the inputs as latent features and then decodes them as the original inputs, an LSTM-RNN predicts the next input from the previous input sequence. Reconstruction errors between observations and the predictions are calculated, and high error patterns are identified as anomalies. Although these methods have achieved good performance, the modeling is based on the acoustic feature space, such as log Mel filterbanks, which cannot model the temporal structure information of the waveform.

To address this issue, a novel anomalous SED method based on the direct waveform modeling using WaveNet [47] was proposed [18, 25]. Because WaveNet is capable of modeling the detailed temporal structures like the phase information, the proposed method can detect anomalous sound events more accurately than conventional methods based on the reconstruction errors of acoustic features [29]. Furthermore, to take differences in environmental situations into consideration, i-vector was used as an additional auxiliary feature of WaveNet, which has been utilized in speaker verification [8]. This i-vector extractor will enable discrimination of the sound patterns, depending on the time, location, and the surrounding environment. Experimental evaluation using a database of sounds recorded in public spaces demonstrated that the proposed method can outperform conventional feature-based anomalous SED approaches and that modeling in the time domain in conjunction with the use of i-vector is effective for anomalous SED.

A detection example of the proposed method is shown in Fig. 8. From the top to bottom, the figures represent the input signal in the time domain, the spectrogram in the time-frequency domain, entropy-based anomalous score, and detected binary

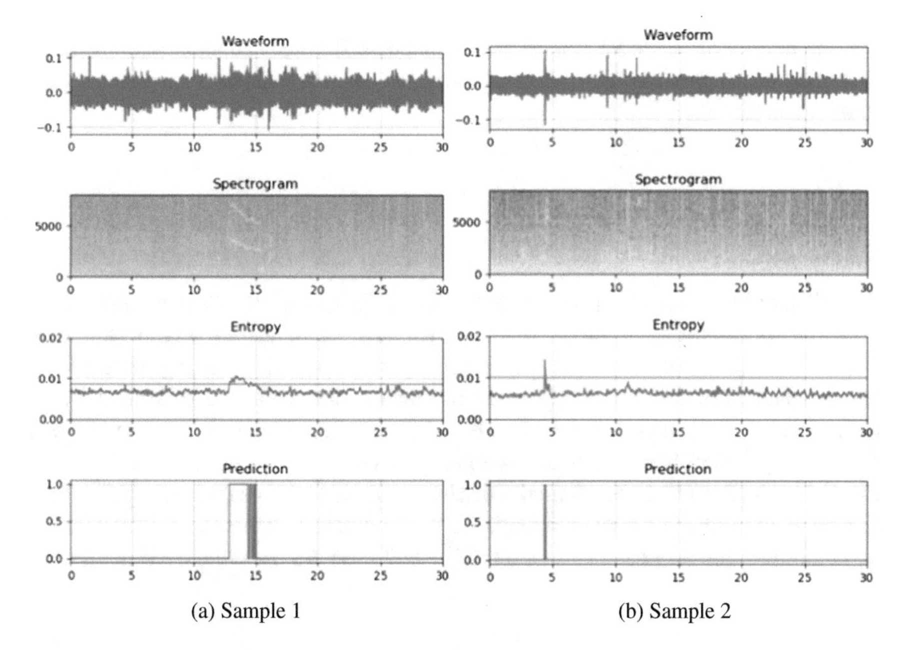

(a) Sample 1 (b) Sample 2

Fig. 8 Detection example of the proposed method. The figures represent the input signal in the time domain, the spectrogram in the time-frequency domain, entropy-based anomalous score, and detected binary results. Only the first pulse corresponds to the anomalous sound events in the spectrogram of the right figure (glass breaking). The other pulses represent normal events (high-heeled footsteps)

results. In the right figure, there are several pulse-like sounds in the spectrogram; however, only the first pulse corresponds to the anomalous sound events (glass breaking). The other pulses represent normal events (high-heeled footsteps). The result shows that the proposed method can detect anomalous sound events precisely even if they are difficult to distinguish on the spectrogram. Consequently, this approach is one of the good examples of the analysis phase in RWDC, focusing on the difference in the modeling domain.

Furthermore, the proposed anomalous SED method can be applied to data acquisition in RWDC. It is necessary to efficiently collect and annotate data to realize practical applications using machine learning techniques. In SED, these annotations include the start and the end timestamps of each sound event and its label. However, since sound events include a wide range of sounds that vary significantly in their acoustic characteristics, duration, and volume, accurate annotation is challenging and costly, even for humans. To address this issue, the proposed anomalous SED model can be used to detect anomalous sound events, i.e., novel sound events, which help to collect the training data without supervised data. Since the novel detected patterns were not included in the training data, it is expected that they could be used as the training data to improve the performance of SED systems. Therefore, the anomalous SED model can be combined with the SED models to construct a data circulation process involving data acquisition, analysis, and feedback to the implementation.

5 Summary

This chapter introduced the problem definition of SED, how the proposed approaches solve the problems to improve the SED performance and the relationship with RWDC. Through the first work on human activity recognition using multi-modal signals, data circulation intended to enhance the quality of life was described. Then, in the second work on polyphonic SED based on duration modeling, a new method of SED was successfully developed by analyzing the characteristics of the acquired data. Finally, in the third work on anomalous SED based on waveform modeling, the cycle of data acquisition, analysis, and feedback to the implementation was demonstrated by combining the proposed anomalous SED method with the other SED methods.

References

1. Adavanne S, Parascandolo G, Pertila P, Heittola T, Virtanen T (2016) Sound event detection in multichannel audio using spatial and harmonic features. In: Processing detection and classification of acoustic scenes and events 2016 workshop, pp 6–10
2. Basseville M, Nikiforov IV (1993) Detection of abrupt changes: theory and application. Prentice-Hall, Inc
3. Cakir E, Heittola T, Huttunen H, Virtanen T (2015) Polyphonic sound event detection using multi label deep neural networks. In: Processing 2015 IEEE international joint conference on neural networks. IEEE, pp 1–7
4. Chen Z, Watanabe S, Erdogan H, Hershey J (2015) Integration of speech enhancement and recognition using long-short term memory recurrent neural network. In: Processing 2015 annual conference of the international speech communication association, pp 3274–3278
5. Choi I, Kwon K, Bae SH, Kim NS (2016) DNN-based sound event detection with exemplar-based approach for noise reduction. In: Processing detection and classification of acoustic scenes and events 2016 workshop, pp 16–19
6. Chung Y, Oh S, Lee J, Park D, Chang HH, Kim S (2013) Automatic detection and recognition of pig wasting diseases using sound data in audio surveillance systems. Sensors 13(10):12929–12942
7. Clavel C, Ehrette T, Richard G (2005) (2005) Events detection for an audio-based surveillance system. In: Processing IEEE international conference on multimedia and expo. IEEE, pp 1306–1309
8. Dehak N, Kenny PJ, Dehak R, Dumouchel P, Ouellet P (2011) Front-end factor analysis for speaker verification. IEEE Trans Audio Speech Lang Proc 19(4):788–798
9. Dessein A, Cont A, Lemaitre G (2013) Real-time detection of overlapping sound events with non-negative matrix factorization. In: Matrix information geometry. Springer, pp 341–371
10. Espi M, Fujimoto M, Kinoshita K, Nakatani T (2015) Exploiting spectro-temporal locality in deep learning based acoustic event detection. EURASIP J Audio Speech Music Proc 1:1–26

11. Eyben F, Böck S, Schuller B, Graves A et al (2010) Universal onset detection with bidirectional long short-term memory neural networks. In: Processing 2010 international society for music information retrieval conference, pp 589–594
12. Foote J (2000) Automatic audio segmentation using a measure of audio novelty. In: Processing 2000 IEEE international conference on multimedia and expo. IEEE, pp 452–455
13. Gemmeke JF, Ellis DP, Freedman D, Jansen A, Lawrence W, Moore RC, Plakal M, Ritter M (2017) Audio set: an ontology and human-labeled dataset for audio events. In: Processing 2017 IEEE international conference on acoustics, speech and signal processing. IEEE, pp 776–780
14. Gers FA, Schmidhuber J, Cummins F (2000) Learning to forget: continual prediction with LSTM. Neural Comput 12(10):2451–2471
15. Graves A, Mohamed AR, Hinton G (2013) Speech recognition with deep recurrent neural networks. In: Processing 2013 IEEE international conference on acoustics, speech and signal processing. IEEE, pp 6645–6649
16. Hayashi T, Nishida M, Kitaoka N, Takeda K (2015) Daily activity recognition based on DNN using environmental sound and acceleration signals. In: Processing 2015 IEEE european signal processing conference. IEEE, pp 2306–2310
17. Hayashi T, Watanabe S, Toda T, Hori T, Le Roux J, Takeda K (2017) Duration-controlled LSTM for polyphonic sound event detection. IEEE/ACM Trans Audio Speech Lang Proc 25(11):2059–2070
18. Hayashi T, Komatsu T, Kondo R, Toda T, Takeda K (2018a) Anomalous sound event detection based on wavenet. In: Processing 2018 IEEE european signal processing conference. IEEE, pp 2494–2498
19. Hayashi T, Nishida M, Kitaoka N, Toda T, Takeda K (2018b) Daily activity recognition with large-scaled real-life recording datasets based on deep neural network using multi-modal signals. IEICE Trans Fund Elect Commun Comput Sci 101(1):199–210
20. Heittola T, Mesaros A, Virtanen T, Eronen A (2011) Sound event detection in multisource environments using source separation. In: Processing 2011 workshop on machine listening in multisource environments, pp 36–40
21. Heittola T, Mesaros A, Eronen A, Virtanen T (2013) Context-dependent sound event detection. EURASIP J Audio Speech Music Proc 1:1–13
22. Hochreiter S, Schmidhuber J (1997) Long short-term memory. Neural Comput 9(8):1735–1780
23. Innami S, Kasai H (2012) NMF-based environmental sound source separation using time-variant gain features. Comput Math Appl 64(5):1333–1342
24. Komatsu T, Toizumi T, Kondo R, Senda Y (2016) Acoustic event detection method using semi-supervised non-negative matrix factorization with mixtures of local dictionaries. In: Processing detection and classification of acoustic scenes and events 2016 workshop, pp 45–49
25. Komatsu T, Hayashiy T, Kondo R, Todaz T, Takeday K (2019) Scene-dependent anomalous acoustic-event detection based on conditional Wavenet and i-vector. In: Processing 2019 IEEE international conference on acoustics, speech and signal processing. IEEE, pp 870–874
26. Liu S, Yamada M, Collier N, Sugiyama M (2013) Change-point detection in time-series data by relative density-ratio estimation. Neural Netw 43:72–83
27. Ma J, Perkins S (2003) Time-series novelty detection using one-class support vector machines. In: Processing 2003 IEEE international joint conference on neural networks, vol 3. IEEE, pp 1741–1745
28. Malhotra P, Vig L, Shroff G, Agarwal P (2015) (2015) Long short term memory networks for anomaly detection in time series. In: Processing european symposium on artificial neural networks, computational intelligence and machine learning. Presses universitaires de Louvain, pp 89–94
29. Marchi E, Vesperini F, Eyben F, Squartini S, Schuller B (2015) A novel approach for automatic acoustic novelty detection using a denoising autoencoder with bidirectional LSTM neural networks. In: Processing 2015 IEEE international conference on acoustics, speech and signal processing. IEEE, pp 1996–2000
30. Markou M, Singh S (2003) Novelty detection: a review Part 1: statistical approaches. Signal Proc 83(12):2481–2497

31. Mesaros A, Heittola T, Benetos E, Foster P, Lagrange M, Virtanen T, Plumbley MD (2017) Detection and classification of acoustic scenes and events: outcome of the DCASE 2016 challenge. IEEE/ACM Trans Audio Speech Lang Proc 26(2):379–393
32. Ochiai T, Matsuda S, Watanabe H, Lu X, Hori C, Katagiri S (2015) Speaker adaptive training for deep neural networks embedding linear transformation networks. In: Processing 2015 IEEE international conference on acoustics, speech and signal processing. IEEE, pp 4605–4609
33. Ono Y, Onishi Y, Koshinaka T, Takata S, Hoshuyama O (2013) Anomaly detection of motors with feature emphasis using only normal sounds. In: Processing 2013 IEEE international conference on acoustics, speech and signal processing. IEEE, pp 2800–2804
34. Over P, Awad G, Michel M, Fiscus J, Sanders G, Shaw B, Kraaij W, Smeaton AF, Quéot G (2013) TRECVID 2012 – An overview of the goals, tasks, data, evaluation mechanisms and metrics. In: Processing 2012 TREC video retrieval evaluation notebook papers and slides, pp 1–58
35. Parascandolo G, Huttunen H, Virtanen T (2016) Recurrent neural networks for polyphonic sound event detection in real life recordings. In: Processing 2016 IEEE international conference on acoustics, speech and signal processing. IEEE, pp 6440–6444
36. Peng YT, Lin CY, Sun MT, Tsai KC (2009) Healthcare audio event classification using hidden Markov models and hierarchical hidden Markov models. In: Processing 2009 IEEE international conference on multimedia and expo. IEEE, pp 1218–1221
37. Sak H, Senior AW, Beaufays F (2014) Long short-term memory recurrent neural network architectures for large scale acoustic modeling. In: Processing 2014 annual conference of the international speech communication association, pp 338–342
38. Sakurada M, Yairi T (2014) (2014) Anomaly detection using autoencoders with nonlinear dimensionality reduction. In: Processing ACM workshop on machine learning for sensory data analysis. ACM, pp 4–11
39. Salamon J, Jacoby C, Bello JP (2014) A dataset and taxonomy for urban sound research. In: Processing 2014 ACM international conference on multimedia. ACM, pp 1041–1044
40. Scott DW (2004) Outlier detection and clustering by partial mixture modeling. In: Processing computational statistics. Springer, pp 453–464
41. Stiefelhagen R, Bernardin K, Bowers R, Rose RT, Michel M, Garofolo J (2008) The CLEAR 2007 evaluation. In: Multimodal technologies for perception of humans. Springer, pp 3–34
42. Stork JA, Spinello L, Silva J, Arras KO (2012) (2012) Audio-based human activity recognition using non-Markovian ensemble voting. In: Processing IEEE international symposium on robot and human interactive communication. IEEE, pp 509–514
43. Tarassenko L, Hayton P, Cerneaz N, Brady M (1995) Novelty detection for the identification of masses in mammograms. In: Processing 1995 international conference on artificial neural networks, pp 442–447
44. Valenzise G, Gerosa L, Tagliasacchi M, Antonacci F, Sarti A (2007) Scream and gunshot detection and localization for audio-surveillance systems. In: Processing 2007 IEEE conference on advanced video and signal based surveillance. IEEE, pp 21–26
45. Valtchev D, Frankov I (2002) Service gateway architecture for a smart home. IEEE Commun Mag 40(4):126–132
46. Wang Y, Neves L, Metze F (2016) Audio-based multimedia event detection using deep recurrent neural networks. In: Processing 2016 IEEE international conference on acoustics, speech and signal processing. IEEE, pp 2742–2746
47. van den Oord A, Dieleman S, Zen H, Simonyan K, Vinyals O, Graves A, Kalchbrenner N, Senior AW, Kavukcuoglu K (2016) WaveNet: a generative model for raw audio. arXiv:160903499
48. Xu M, Xu C, Duan L, Jin JS, Luo S (2008) Audio keywords generation for sports video analysis. ACM Trans Mult Comput Commun Appl 4(2):1–23
49. Yu SZ (2010) Hidden semi-Markov models. Artif Intel 174(2):215–243

A Study on Utilization of Prior Knowledge in Underdetermined Source Separation and Its Application

Shogo Seki

Abstract In the field of environmental sound recognition, source separation is one of the core technologies used to extract individual sound sources from mixed signals. Source separation is closely related to other acoustic technologies and is used to develop various applications such as automatic transcription systems for meetings, active music listening systems, and music arranging systems for composers. When a mixed signal is composed of more sources than the number of microphones, i.e., in an underdetermined source separation scenario, separation performance is still limited and there remains much room for improvement. Moreover, depending on the method used to extract the source signals, subsequent systems using the acoustic features calculated from the estimated source information can suffer from performance degradation. Supervised learning is a promising method which can be used to alleviate these problems. Training data composed of source signals, as well as mixed signals, is used to obtain as much prior information about the sound sources as possible into account. Supervised learning is essential for improving the performance of underdetermined source separation, although there are problems which remain to be addressed. This study addresses two problems with the supervised learning approach for underdetermined source separation and its application. The first is how to improve the use of prior information, and the second is how to improve the representation ability of source models. To deal with the first problem, (1) the characteristics of individual source signals in the spectral and feature domains and (2) the temporal characteristics implicitly considered in time-frequency analysis are focused. Furthermore, this study also explores the use of deep generative models for prior information to deal with the second problem.

S. Seki (✉)
Graduate School of Informatics, Nagoya University, Nagoya, Japan
e-mail: seki.shogo@g.sp.m.is.nagoya-u.ac.jp

K. Takeda et al. (eds.), *Frontiers of Digital Transformation*,
https://doi.org/10.1007/978-981-15-1358-9_3

1 Introduction

As depicted in Fig. 1, human beings can perceive where sound signals come from, discriminate and identify various sounds, and recognize related events or meanings. A number of studies have attempted to replicate these complicated functions of human beings using technology, as part of a field of study known as Computational Auditory Scene Analysis (CASA) [4, 5]. Recently, researchers have been focusing on environmental sounds, as well as on speech and music.

Fig. 1 Overview of the hearing abilities of human beings and the research field of computational auditory scene analysis. Source localization (solid black), source separation (dashed black), and automatic speech recognition (dotted black) refer to the problems of determining the allocations of sound sources, identifying the individual sound sources from an observed mixture signal, and recognizing each sound signal as text, which has been studied for a long time and have mainly focused on speech and acoustic sounds. Acoustic scene classification (solid gray) and acoustic event detection (dashed gray) refer to the problems of categorizing the situations of surrounded environments and spotting the acoustic events in recorded signals, respectively, which deal with environmental signals and are recently receiving more attention

Source separation [9] is one of the most important CASA technologies, which allows the separation of individual source signals from a mixture of signals. Although humans can easily pick out various sounds from a noisy background, such as birds chirping, people talking, traffic noises, rain falling, etc., this is a very difficult task to automate. Source separation is used in conjunction with other technologies to develop various applications. By combining source separation with Automatic Speech Recognition (ASR) [8], we can develop systems that can automatically identify different speakers in a meeting and transcribe each person's utterances. By combining source separation with binaural techniques [23], it is possible to develop new sound systems that allow music listeners to adjust a wide variety of variables in order to achieve their favorite allocation of sound sources, i.e., active music listening systems. Using source separation with Voice Conversion (VC) [1] could potentially allow composers to extract the vocal components of a music signal and convert them by giving them different or additional attributes.

For decades, source separation has been studied and developed under blind conditions, i.e., Blind Source Separation (BSS) [7, 9, 22], where no information about the number or location of source signals, or the mixing process is given. Categorization of source separation problems is shown in Fig. 2. If we can use the same number of microphones as the number of source signals, or even a larger number of microphones, which are called determined and overdetermined source separation, respectively, impressive source separation performance can be achieved [6, 12, 15, 18, 25]. In contrast, when a mixture of signals is composed of more sources than the number of microphones, which is called underdetermined source separation [26, 28], separation performance is still limited, so there is room for improvement.

Supervised learning [2, 24], which uses training data composed of the source signals contained within the mixed signals, is a promising way to alleviate this problem, by taking into account as much information about the sources, i.e., prior information, as possible. The source estimation method which is used is important because subsequent processing can be impaired if the source estimation data are inaccurate. Time-frequency masking [29] is one example of such a source estimation approach. Since non-target components are over-suppressed and target components remain only sparsely, the acoustic features calculated from the masked sources degrade during subsequent processing.

This study addresses two problems which are encountered when using a supervised approach for underdetermined source separation. One is how to utilize prior information more efficiently, and the other is how to improve the representation ability of a model when using prior information.

Synthesized music signals, such as the music distributed on CDs or through online music websites, are generally stereophonic signals composed of linear combinations of many individual source signals and their mixing gains, in which spatial information, i.e., phase information, or its differential between each channel cannot be utilized as acoustic clues for source separation. To separate these stereophonic music signals, this study employs the concept of Non-negative Matrix Factorization (NMF) [19, 27] and proposes a supervised source separation method based on Non-negative Tensor Factorization (NTF) [3], a multi-dimensional extension of NMF. In order to reflect

Overdetermined condition **Determined condition**

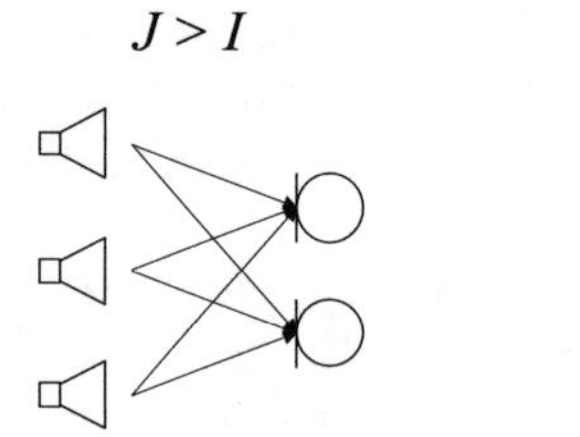

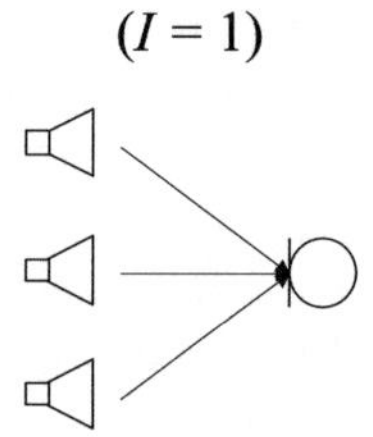

Underdetermined condition

Fig. 2 Categorization of source separation, which depends on the relationship between the number of source signals J and that of microphones I. Overdetermined source separation and determined source separation refer to the problems of estimating source signals using a mixture signal obtained by microphones, where the number of source signals is less than or equal to that of microphones. Using the limited number of microphones, underdetermined source separation attempts to estimate the larger number of source signals, which includes single-channel source separation as an ultimate case

prior information of each source efficiently, this study further introduces the Cepstral Distance Regularization (CDR) [20] method to consider and regularize the timbre information of the sources. Experimental results show that CDR yields significant improvement in separation performance and provides better estimation for mixing gains. Experimental results show that CDR yields significant performance improvements and provides better estimation for mixing gains.

After estimating the source signals in a mixed signal, time-frequency masking is a well-known approach used to extract source signals for source separation and speech enhancement [21]. While it is very effective in terms of signal recovery accuracy, e.g., signal-to-noise ratio, one drawback is that it can over-suppress and damage speech components, resulting in limited performance when used with succeeding speech processing systems. To overcome this flaw, this study proposes a method to restore the missing components of time-frequency masked speech spectrograms, which is based on direct estimation of a time-domain signal, referred to as Time-domain Spectrogram Factorization (TSF) [10, 11]. TSF-based missing component restoration allows us to take into account the local interdependencies of the elements of complex spectrograms derived from the redundancy of a time-frequency representation, as well as the global structure of the magnitude spectrogram. Experimental results demonstrate that the proposed TSF-based method significantly outperforms conventional methods and has the potential to estimate both phase and magnitude spectra simultaneously and precisely.

When solving underdetermined source separation problems, Multichannel Non-negative Matrix Factorization (MNMF) [26, 28], a multichannel extension of NMF, adopts the NMF concept to model and estimate the power spectrograms of the sound sources in a mixed signal. Although MNMF works reasonably well for particular types of sound sources, it can fail to work for sources with spectrograms that do not comply with NMF, resulting in limited performance. However, a supervised source separation method called a Multichannel Variational AutoEncoder (MVAE) [13, 14], which is an improved variant of determined source separation methods, has been proposed, in which a conditioned Variational AutoEncoder (VAE) [16], i.e., conditional VAE [17] is used instead of the NMF model for modeling source power spectrograms. This chapter proposes a generalized method of MVAE called the GMVAE method, which is constructed by combining the features of an MNMF and an MVAE so that it can also deal with underdetermined source separation problems. Experimental evaluations demonstrate that GMVAE outperforms baseline methods including MNMF.

This study is organized as follows: In Sect. 2, stereophonic music source separation using the timbre information of sources is described. In Sect. 3, missing component restoration by considering the redundancy of time-frequency representation is described. In Sect. 4, multichannel source separation based on a deep generative model is described. In Sect. 5, the relationship between source separation and Real-World Data Circulation (RWDC) is mentioned. In Sect. 6, the contributions of this study and future work are discussed.

2 Stereophonic Music Source Separation Using Timbre Information of Sources

This section proposes a supervised source separation method for stereophonic music signals containing multiple recorded or processed signals, where synthesized music is focused as the stereophonic music. As the synthesized music signals are often generated as linear combinations of many individual source signals and their respective mixing gains, phase or phase difference information between inter-channel signals, which represent spatial characteristics of recording environments, cannot be utilized as acoustic clues for source separation. Non-negative Tensor Factorization (NTF) is an effective technique which can be used to resolve this problem by decomposing amplitude spectrograms of stereo channel music signals into basis vectors and activations of individual music source signals, along with their corresponding mixing gains. However, it is difficult to achieve sufficient separation performance using this method alone, as the acoustic clues available for separation are limited.

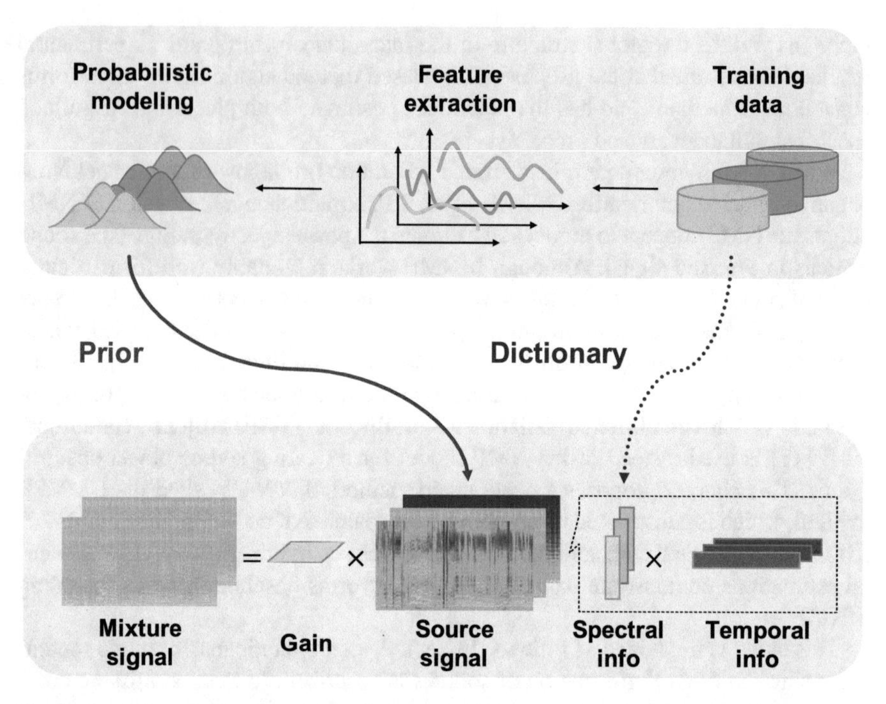

Fig. 3 Overview of the proposed stereophonic music source separation method. Conventional source separation approaches use an external dataset for the source signals present in a mixture signal to train dictionaries representing spectral information of each source signal (dotted). The proposed method utilizes the dataset not only to train the dictionaries but also to build their probabilistic models in the feature domain, which are a priori (solid) and regularizes the source signals in the source estimation

To address this issue, this section proposes a Cepstral Distance Regularization (CDR) method for NTF-based stereo channel separation, which involves making the cepstrum of the separated source signals follow Gaussian Mixture Models (GMMs) of the corresponding music source signal. Figure 3 shows the overview of the proposed method with a conventional source separation method. These GMMs are trained in advance using available samples. Experimental evaluations separating three and four sound sources are conducted to investigate the effectiveness of the proposed method in both supervised and partially supervised separation frameworks, and performance is also compared with that of a conventional NTF method. Experimental results demonstrate that the proposed method yields significant improvements within both separation frameworks, and that CDR provides better separation parameters.

3 Missing Component Restoration Considering Redundancy of Time-Frequency Representation

While time-frequency masking is a powerful approach for speech enhancement in terms of signal recovery accuracy, e.g., signal-to-noise ratio, it can over-suppress and damage speech components, leading to limited performance of succeeding speech processing systems (Fig. 4). To overcome this shortcoming, this section proposes a method to restore missing components of time-frequency masked speech spectrograms based on direct estimation of a time-domain signal. The overview of the proposed method is depicted in Fig. 5. The proposed method allows us to take into account the local interdependencies of the elements of the complex spectrogram derived from the redundancy of a time-frequency representation as well as the global structure of the magnitude spectrogram. The effectiveness of the proposed method is demonstrated through experimental evaluation, using spectrograms filtered with masks to enhance noisy speech. Experimental results show that the proposed method significantly outperforms conventional methods, and has the potential to estimate both phase and magnitude spectra simultaneously and precisely.

4 Multichannel Source Separation Based on a Deep Generative Model

This section discusses the multichannel audio source separation problem under underdetermined conditions. MNMF is a powerful method for underdetermined audio source separation, which adopts the NMF concept to model and estimate the power spectrograms of the sound sources in a mixture signal. This concept is also used in Independent Low-Rank Matrix Analysis (ILRMA), a special class of the MNMF formulated under determined conditions. While these methods work reasonably well for particular types of sound sources, one limitation is that they can fail to work for sources with spectrograms that do not comply with the NMF model.

To address this limitation, an extension of ILRMA called the Multichannel Variational AutoEncoder (MVAE) method was recently proposed, where a Conditional VAE (CVAE) is used instead of the NMF model for expressing source power spectrograms. This approach has performed impressively in determined source separation tasks thanks to the representation power of deep neural networks.

While the original MVAE method was formulated under determined mixing conditions, a generalized version of it by combining the ideas of MNMF and MVAE is proposed so that it can also deal with underdetermined cases. This method is called the Generalized MVAE (GMVAE) method. Figure 6 shows a block diagram of the proposed GMVAE with the conventional MNMF. Source signals estimated by the conventional method and the proposed method are shown in Fig. 7. When compar-

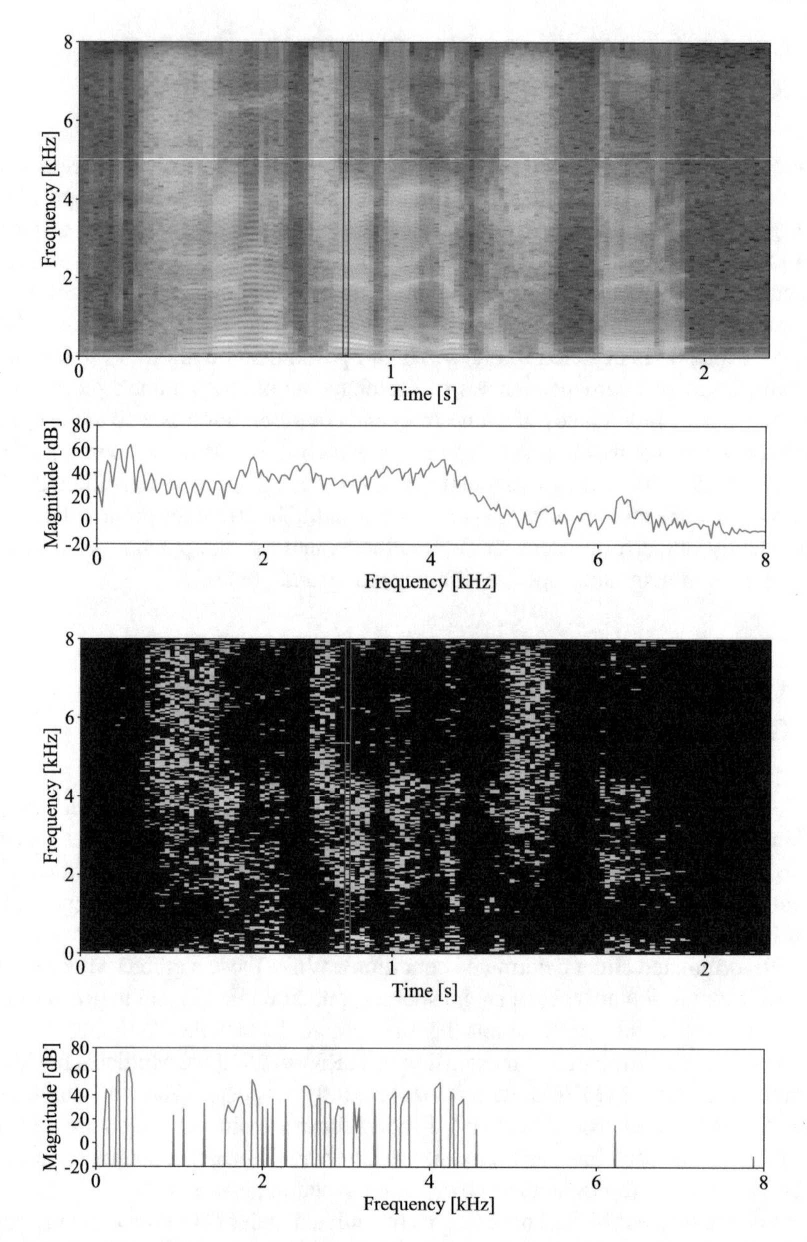

Fig. 4 Comparison of clean speech (top) and estimated speech obtained by masking time-frequency components of noisy speech (bottom), where the acoustic features used in succeeding systems are calculated from each spectrum

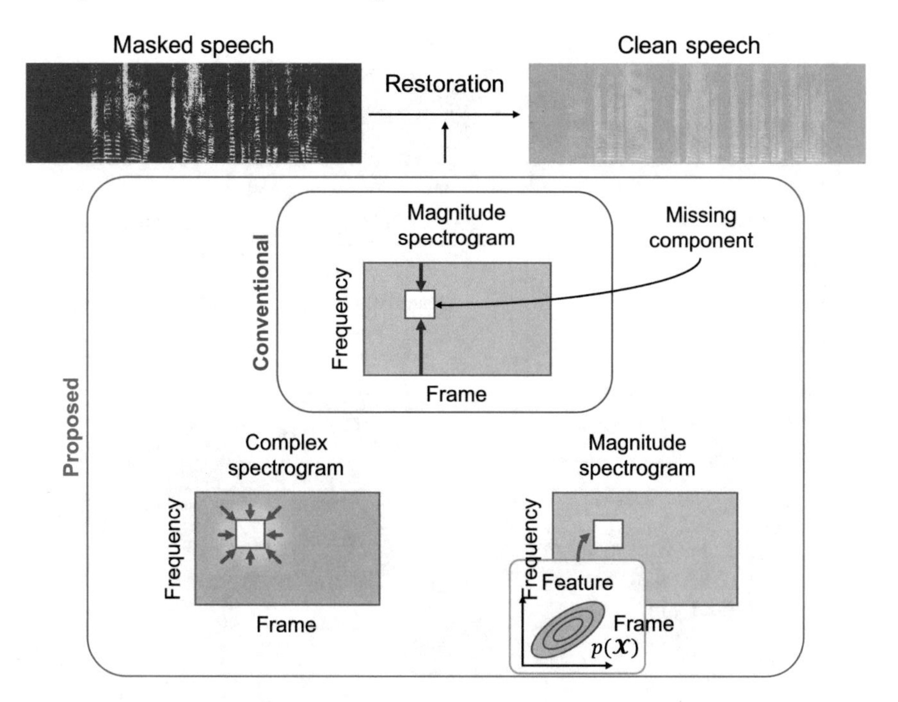

Fig. 5 Overview of the proposed missing component restoration method. Missing component restoration refers to the problem of replacing the lacking components of masked speech with substituted values and recovering clean speech. The conventional method only takes the fact that the time-frequency representation of a speech signal can form a low-rank structure in the magnitude domain into account, which leads to the estimation and the replacement of the missing component by considering only the global structure of a masked speech. The proposed method considers the local interdependencies of each time-frequency component in the complex domain and prior information in the feature domain as well as the global structure in the magnitude domain considered in the conventional method

ing the estimated signals Fig. 7(b) and Fig. 7(c), the proposed method represent the reference signal Fig. 7(a) more accurately. In underdetermined source separation and speech enhancement experiments, the proposed method performs better than baseline methods.

5 Relationship with RWDC

Real-World Data Circulation (RWDC) is a multidisciplinary study on the flow of acquisition, analysis, and implementation data, and the circulation of this flow, which is a key to the successful development of commercial services and products. In the third section, the analysis and the application flow in RWDC are reviewed, and

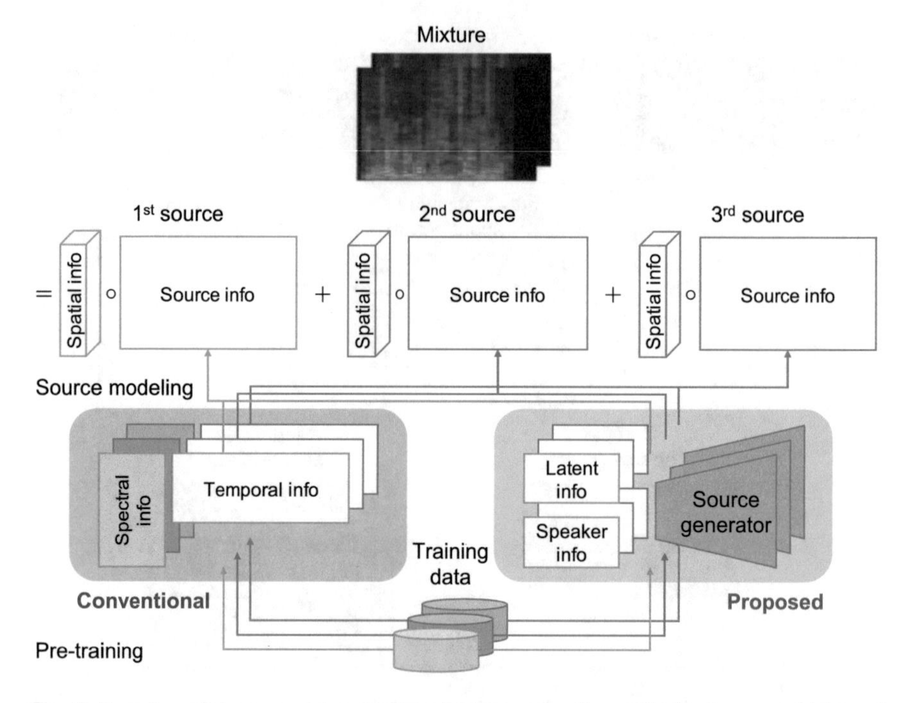

Fig. 6 Overview of the proposed multichannel source separation method, where a multichannel mixture signal is supposed to consist of three source signals with each of them modeled by the combination of spatial information and source information. For supervised source separation, a conventional multichannel source separation method uses training data for each source signal to train spectral information (dictionaries) with the corresponding temporal information to represent source information. The learned spectral dictionaries are then used as prior information for the target mixture signal. One of the shortcomings of the conventional method is that the representation capabilities are limited due to the linear combination of spectral and temporal information. The proposed method pre-trains a universal deep generative model that generates an arbitrary speech of multiple speakers from latent information and speaker information, and utilizes the trained source generator to represent source information, where the representation capabilities are high thanks to the use of a neural network

describes how source separation problems and RWDC are related are discussed. In addition, how source separation can contribute to creating new value is discussed.

The development of new commercial services and products begins with the collection of real-world data about the needs and desires of potential customers. This data is analyzed and the output is used to refine these new services and products. Once launched, feedbacks from customers about the new applications are also collected, and thus there should be a certain flow of data circulation during each phase of development. RWDC is a multidisciplinary study that considers the flow of acquisition, analysis, and implementation of data, as well as the circulation of this flow.

In the data acquisition phase, various phenomena in the real world are acquired in the form of digital data through observations. In the data analysis phase, information

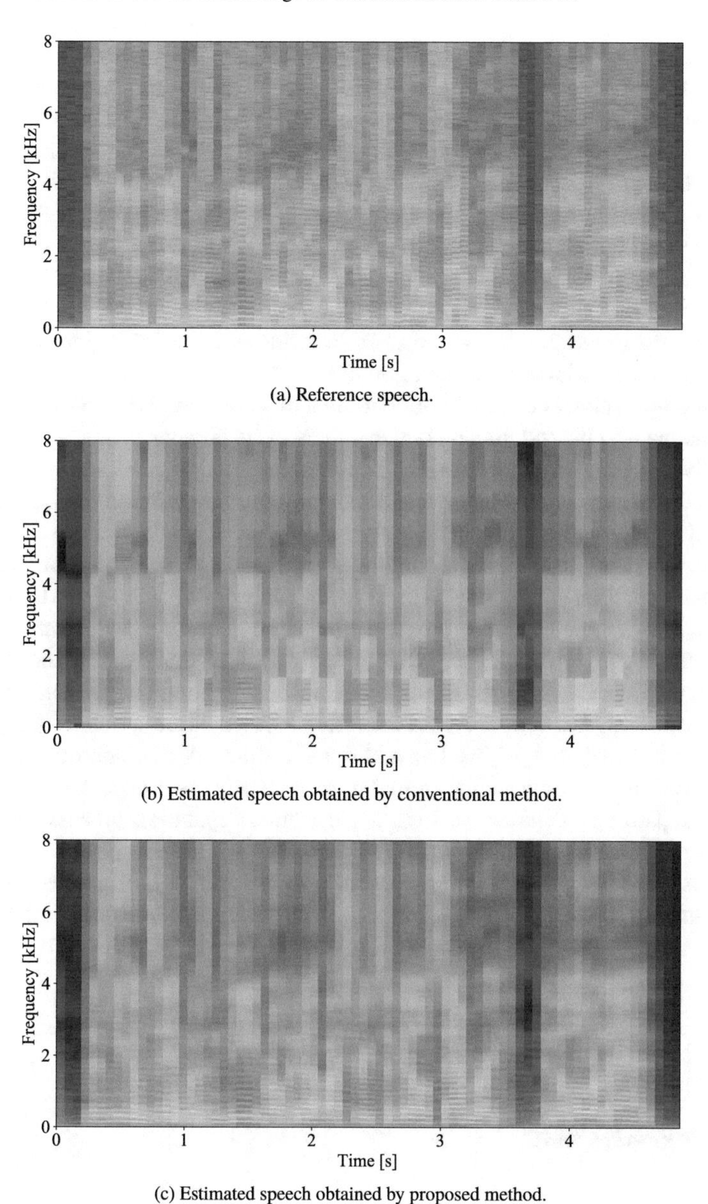

Fig. 7 Power spectrograms of reference speech (a) and estimated speech obtained by conventional method (b) and the proposed method (c)

technologies such as pattern recognition and machine learning techniques are used to reveal the characteristic or structure of the data. In the data implementation phase, the output results of the data analysis are used to create new services and products. Thus, iteratively repeating this flow, new services and products can be developed which reflect users' demands, creating a new social value.

From a value creation point of view, two connections between RWDC and the source separation techniques proposed in this study are discussed in this section. One is data circulation within the source separation problem, and the other is source separation during data circulation. What kind of data circulation can be expected in source separation problems, and also, in terms of data circulation, how source separation can contribute to the creation of new value are discussed.

Source separation is a method of dividing a mixture of audio signals recorded by microphones into the individual source signals of its components, a process which is closely related to data circulation. The method of supervised source separation addressed in this study is an improved method that uses training data to obtain as much prior information into account about the source signals in a mixed signal as possible, which results in better separation performance. This use of prior knowledge for signal separation can be viewed as a type of analysis of the input mixture signals. The outputs of source separation are the estimated source signals, which are generally used in subsequent systems. The goal is to transmit clean signal data to the following systems, and this flow of processed data represents the implementation phase of a data circulation system. Moreover, if the estimated signals are sufficiently separated, these output signals can be used as additional training data. Consequently, due to this increase in training data, the source separation process is expected to be further improved. Thus, as depicted in Fig. 8, the source separation process itself is an example of data circulation.

As described in the previous section, source separation has been studied and developed in order to replicate complex human hearing functions. As a result, it has a great

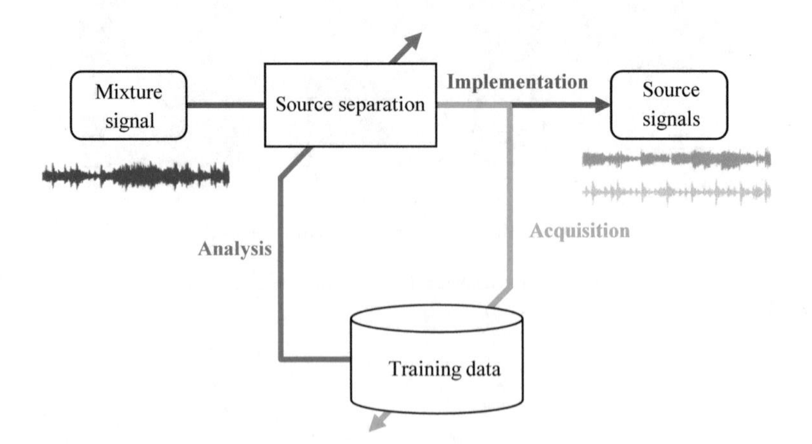

Fig. 8 Data circulation in (supervised) source separation

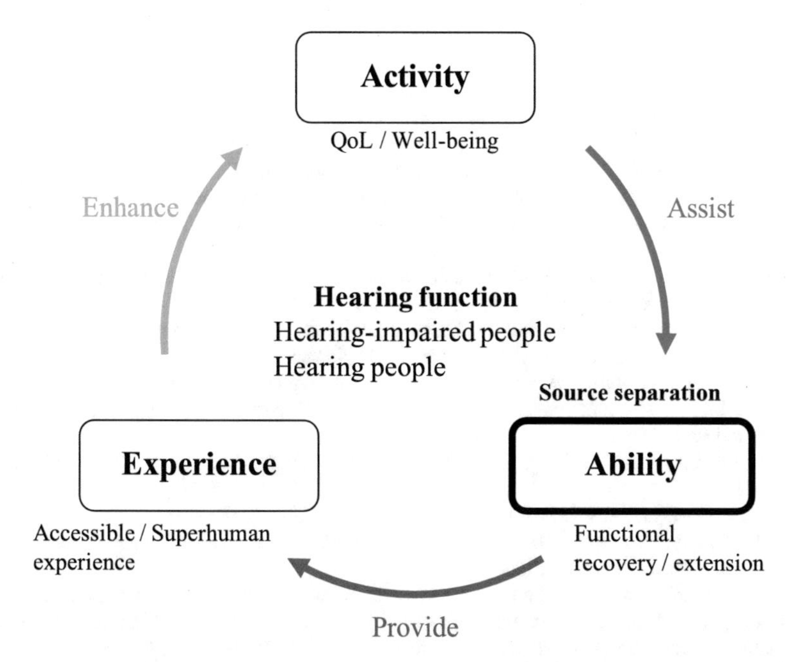

Fig. 9 Value creation through source separation. Separation of audio signal data can improve the hearing abilities of the public

potential for replacing or augmenting human hearing. Figure 9 illustrates the cyclic flow of human hearing activity being improved through the use of source separation. The output separated signals can be used in various kinds of hearing enhancement applications, allowing people to acquire additional abilities, i.e., functional recovery for hearing-impaired persons or functional extension for people with normal hearing. Thanks to their improved abilities, users of these applications can enjoy new experiences, e.g., improved hearing allows people with impaired hearing to function more normally, while enhanced hearing for unimpaired users provides superhuman hearing experiences. Ultimately, these experiences enhance the hearing abilities and functioning of each user, which provides the public with improved QoL and a higher level of wellness.

6 Conclusion

I have introduced the concept of RWDC and reviewed the components of RWDC. Data circulation during source separation, and source separation as a component of RWDC processes for providing improved products and services, i.e., how source separation can contribute to value creation, were also discussed.

References

1. Abe M, Nakamura S, Shikano K, Kuwabara H (1990) Voice conversion through vector quantization. J Acoust Soc Japan (E) 11(2):71–76
2. Bishop CM (2006) Pattern recognition and machine learning. Springer
3. Cichocki A, Zdunek R, Phan AH, Amari Si (2009) Nonnegative matrix and tensor factorizations: applications to exploratory multi-way data analysis and blind source separation. John Wiley & Sons
4. Cooke M, Brown GJ (1993) Computational auditory scene analysis: exploiting principles of perceived continuity. Speech Commun 13(3–4):391–399
5. Cooke M, Brown GJ, Crawford M, Green P (1993) Computational auditory scene analysis: Listening to several things at once. Endeavour 17(4):186–190
6. Hiroe A (2006) Solution of permutation problem in frequency domain ica, using multivariate probability density functions. In: International Conference on Independent Component Analysis and Signal Separation (ICA), pp 601–608
7. Hyvärinen A, Karhunen J, Oja E (2004) Independent component analysis, vol 46. John Wiley & Sons
8. Jelinek F (1997) Statistical methods for speech recognition. MIT press
9. Jutten C, Herault J (1991) Blind separation of sources, part i: an adaptive algorithm based on neuromimetic architecture. Signal Proc 24(1):1–10
10. Kagami H, Kameoka H, Yukawa M (2017) A majorization-minimization algorithm with projected gradient updates for time-domain spectrogram factorization. In: IEEE international conference on acoustics, speech and signal processing (ICASSP). IEEE, pp 561–565
11. Kameoka H (2015) Multi-resolution signal decomposition with time-domain spectrogram factorization. In: IEEE international conference on acoustics, speech and signal processing (ICASSP). IEEE, pp 86–90
12. Kameoka H, Yoshioka T, Hamamura M, Le Roux J, Kashino K (2010) Statistical model of speech signals based on composite autoregressive system with application to blind source separation. In: International conference on latent variable analysis and signal separation (LVA/ICA), pp 245–253
13. Kameoka H, Li L, Inoue S, Makino S (2018) Semi-blind source separation with multichannel variational autoencoder. arXiv:180800892
14. Kameoka H, Li L, Inoue S, Makino S (2019) Supervised determined source separation with multichannel variational autoencoder. Neural Comput 31(9):1891–1914
15. Kim T, Eltoft T, Lee TW (2006) Independent vector analysis: an extension of ica to multivariate components. In: International conference on independent component analysis and signal separation (ICA), pp 165–172
16. Kingma DP, Welling M (2014) Auto-encoding variational bayes. In: International conference on learning representations (ICLR)
17. Kingma DP, Mohamed S, Rezende DJ, Welling M (2014) Semi-supervised learning with deep generative models. In: Advances in neural information processing systems, pp 3581–3589
18. Kitamura D, Ono N, Sawada H, Kameoka H, Saruwatari H (2016) Determined blind source separation unifying independent vector analysis and nonnegative matrix factorization. IEEE/ACM Trans Audio Speech Lang Proc 24(9):1622–1637
19. Lee DD, Seung HS (1999) Learning the parts of objects by non-negative matrix factorization. Nature 401(6755):788
20. Li L, Kameoka H, Higuchi T, Saruwatari H (2016) Semi-supervised joint enhancement of spectral and cepstral sequences of noisy speech. In: Interspeech, pp 3753–3757
21. Loizou PC (2013) Speech enhancement: theory and practice. CRC Press
22. Makino S, Lee TW, Sawada H (2007) Blind speech separation, vol 615. Springer
23. Møller H (1992) Fundamentals of binaural technology. Appl Acoust 36(3–4):171–218
24. Murphy KP (2012) Machine learning: a probabilistic perspective. The MIT Press

25. Ono N (2011) Stable and fast update rules for independent vector analysis based on auxiliary function technique. In: IEEE international workshop on applications of signal processing to audio and acoustics (WASPAA), pp 189–192
26. Ozerov A, Févotte C (2010) Multichannel nonnegative matrix factorization in convolutive mixtures for audio source separation. IEEE Trans Audio Speech Lang Proc 18(3):550–563
27. Paatero P, Tapper U (1994) Positive matrix factorization: a non-negative factor model with optimal utilization of error estimates of data values. Environmetrics 5(2):111–126
28. Sawada H, Kameoka H, Araki S, Ueda N (2013) Multichannel extensions of non-negative matrix factorization with complex-valued data. IEEE Trans Audio Speech Lang Proc 21(5):971–982
29. Yilmaz O, Rickard S (2004) Blind separation of speech mixtures via time-frequency masking. IEEE Trans Signal Proc 52(7):1830–1847

A Study on Recognition of Students' Multiple Mental States During Discussion Using Multimodal Data

Shimeng Peng

Abstract Augmenting human teacher's perceptual and reasoning capability by leveraging machine intelligence, in monitoring the multiple mental states of students exposed in discussion activities, inferring the need for latent assistance, has been on the agenda of educational community. The development of modern sensing technologies makes it possible to generate and accumulate massive amounts of multimodal data, which offers opportunities for supporting novel methodological approaches that automatically measure the mental states of students from various perspectives during a multi-participant discussion activity where the dynamic and time-sensitive decision-making plays a major role. In this chapter, an advanced multi-sensor-based data collection system is introduced which is applied in a real university research lab to record and archive a multimodal "in-the-wild" teacher-student conversation-based discussion dataset for a long term. A line of data-driven analysis works is performed to explore how we could transform the raw sensing signals into context-relevant information to improve decision-making. In addition, how to design multimodal analytics to augment the ability to recognize different mental states and the possibility of taking advantages of the supplement and replacement capabilities between different modalities to provide human teachers with solutions for addressing the challenges with monitoring students in different real-world education environments is also presented.

1 Introduction

Academic discussion is one of the most common interactive cognitive learning activities held in higher education today, in which students are required to complete a series of complex cognitive tasks including answering casual questions (Q&A), generating explanations, solving problems, demonstrating and transferring the acquired knowledge to conduct further discussion, etc. [2]. A broad range of remarkable research

---

S. Peng (✉)
Department of Intelligent Systems, Graduate School of Informatics,
Nagoya University, Nagoya, Japan
e-mail: hou@nagao.nuie.nagoya-u.ac.jp

K. Takeda et al. (eds.), *Frontiers of Digital Transformation*,
https://doi.org/10.1007/978-981-15-1358-9_4

work in educational psychology has validated the idea that students may consistently experience a mixture of multiple learning-centered mental states, such as the certainty level, engaged/concentration, confusion, frustration, boredom, etc., in complex cognitive learning and those mental states can be used as crucial components for revealing students' learning situations [3, 8, 9, 11, 32]. An ideal teacher should be sufficiently sensitive to monitor students' "assistance dilemma" states [20] exposed in discussion, such as when students lack the certainty to give correct answers, when a student is found to be confused with the current discussion opinion, or when the student is found to be frustrated with the discussion content, in which students cannot rely on their own ability to resolve those dilemma states in certain learning goals but requires external guidance and intervention from teachers.

For one-to-one coaching activities, observing students' external responses, such as their facial expressions or speech statements, is a common way for teachers to monitor students' learning situations and to determine what kind of support to provide and at which timings [23]. However, for the case of offline multi-participant discussion activities, teachers have difficulty capturing changes in the mental states of each participant, especially for those students who engage in few interactions in a discussion and who, most of the time, prefer to participate as if they were an audience member. On the other hand, carrying out remote lectures or discussion activities has gradually become a popular form of modern coaching; at the same time, because some students tend not to use cameras or tend to mute microphones during remote educational activities, this will lead to facial and auditory cues being completely unavailable at certain timings, which undoubtedly brings another challenge for teachers in capturing the mental state of students.

Many lines of research have explored the automatic measurement of students' single mental state such as in terms of engagement in pre-designed student versus computer-teacher tasks, such as problem-solving, essay writing, programming testing, and game design [11, 14, 18, 29], where some of them use only uni-variate modality signals, that is, video [14], audio [11], and physiological measures [18]. However, it is still an open question as to how to effectively monitor students' multiple mental states, such as certainty level, concentration, confusion, frustration, and boredom, when they are interacting with a human teacher in an offline multi-participant conversation-based discussion, and also how to address the challenges with monitoring students in different educational settings. To cope with these challenges, this research attempts to leverage multiple sensory signals including facial, heart rate, and acoustic signals to design an intelligent monitoring agent that can effectively sense the multiple mental states of students during discussion-based learning, and to explore how to design multimodal analytics to provide human teachers with solutions to augment their perceptual capabilities to address the challenges of monitoring students in different real-world educational environments.

Novelty and Contribution

There are several novel contributions in this research from aspects that are preliminarily different from relevant studies: (1) A multi-sensor-based data collection system is developed for supporting the generation and accumulation of massive amounts of multimodal data in "in-the-wild" educational settings, since "in-the-wild" contexts with real operational environments and real teacher-student conversations pose unique challenges in terms of collecting, validating, and interpreting data. This will allow us to provide an enormous amount of rich high-frequency data resources to support other multi-angle analysis work in real-world educational activities. (2) Instead of learning interactions between students and computer tutors or pre-designed script-based learning activities in both HCI and HHI environments, this research focuses more in paying attention to an "un-plugged" scenario in which students and their advisor teacher have a coaching-driven conversation in real discussion-based learning activities. Simply put, the study aims to analyze a series of "true feelings" exposed during these real conversations, which guarantee the results provide evidence of the potential practical value of fusing those multimodal data to explore students' "in-the-wild" learning-centered mental states, increasing the applicability and practicality of the results for real-world coaching activities. (3) With few exceptions, most existing work has focused on using a uni-variate modality to analyze students' single mental state, such as engagement, or basic emotional states such as joy and sadness. In comparison, this research attempts to integrate multiple modalities, facial, heart rate, and acoustic cues to generate an intelligent monitoring agent that effectively senses multiple learning-centered mental states of students, that is, certainty level, concentration, confusion, frustration, and boredom.

2 Related Work

Video-Based Modality

With the development of computer vision technologies, there has been a rich body of research work that uses facial features extracted from video streams for the task of detecting human mental states. Hoque et al. derived a set of mouth-related features from video to characterize smiling movements and explored the possibility of a "smile" being used to identify frustration or delight [17]. Gomes et al. employed eye-related features from facial signals like blinking and gaze to analyze students' concentration states during learning activities [12]. Grafsgaard et al. used a series of video-based cues to characterize facial expressions and predicted students' engagement, frustration, and learning gain [13]. Bosch et al. used several facial-related features extracted from video to describe the movement in the brow, eye, and lip areas, and they trained several supervised learning models to predict multiple mental states when students are playing a physics game in a computer environment [1].

Physiological-Based Modality

More recent work in this space has been able to accurately predict students' learning-centered mental states or simply basic emotional states when they are engaged in learning activities. Hellhammer assessed HR changes before, during, and after cognitive tasks to measure students' stress level [16]. Pereira et al. used HR and HRV to predict students' stress state [30]. In the research of Muthukrishnan et al., they validated the predictive ability of HRV features in predicting learning performance [27].

Audio-Based Modality
It is widely believed that emotion/mental state information may be transmitted from speech signals and can be explicated from linguistic and audio channels. Reilly et al. used a set of linguistic features to predict how students reached consensus and their level of understanding of problems [31]. Kovanovic et al. took advantage of text mining technologies to extract a number of text-based features from online discussion transcripts to predict students' cognitive presence [21]. In addition, Castellano et al. proposed a method for extracting speech features including MFCC, pitch contour, etc. with other modality cues to classify eight basic emotions: anger, despair, interest, irritation, joy, pleasure, pride, and sadness [4].

Multimodal Learning Analytics
Most recently, with the emergence of modern sensors that provide support for recording Multimodal Data (MMD) of students in learning activities, it is now possible to use novel methodological approaches to measure students' multiple affective/mental states from various perspectives, and this has been explored to improve recognition accuracy. Monkaresi et al. used facial cues and heart rate cues to predict student engagement as they work on writing tasks in a computer environment [26]. Peng et al. integrated multiple modalities of facial and EEG signals from a group of middle school students to describe their engagement level when they interacted with an online learning tutor system [29].

3 Participants and Study Protocol

Two progressive experimental investigations were conducted to characterize and predict several graduate students' multiple mental states when they are having conversation-based discussion with their advisor professor in a real university's lab. The students ranged in age from 22 to 25 years. The professor has been guiding these students for 2–3 years by holding a regular real-lab seminar discussion every week. For each discussion, one presenter-student reports a research topic while displaying slides, and completes a number of Q&A session and further discussion session with the meeting participants including the presenter-students' peers and their advisor professor.

To start with an exploitative study was conducted to verify the first argument that if a student's physiological cue (heart rate) can be used as an effective predictor to infer their answer correctness of Q&A session, along with exploring different reasoning abilities shown by heart rate cues and traditional text-based cues in predicting the answer correctness in Q&A sessions. Next, the research scope is extended to the entire conversation-based discussion, and facial and audio information along with heart rate data are adopted to characterize students' multiple mental states including concentration, confusion, frustration, and boredom from various aspects, and how to design multimodal analytics to generate an optimal intelligent monitoring agent of multiple learning-centered mental states of students are explored.

4 Study I: Exploitative Study on Automatic Detection of Students' Answer Correctness in Q&A Based on Their Heart Rate Cues

4.1 Data Collection Methodology

The Health Kit framework running on Apple Watch is used to monitor and collect presenter-student's real-time heart rate data, with a frequency of 1.0 Hz, during the whole discussion, as shown in Fig. 1. Audio-video information of discussion meetings was recorded by a semi-automatic Discussion-Mining (DM) system [28] which was developed and used in our laboratory to record audio-visual and semantic information of Q&A sessions given by participants and answered by the discussion presenter-students, as shown in Fig. 1.

A web-based concurrent annotation tool was developed and installed in tablets which were handled by each participant during the meeting to collect the evaluation of the answer's appropriateness as ground-truth data. A total of 247 Q&A segments of 11 students, with an average of around 22 Q&A segments answered by each student, with a mean length of approximately 5 min per Q&A segment were recorded. Those 247 Q&A segments were annotated concurrently by discussion participants immediately after each question was answered by presenter-students based on a five-point scale. Then the annotation was treated as a binary labeling task into appropriate or inappropriate, where 112 Q&A segments were evaluated as inappropriate and 135 Q&A segments were evaluated as appropriate. Cohen's Kappa value [22] was used to measure the inter-rater agreement between the question-askers and advisor professor.

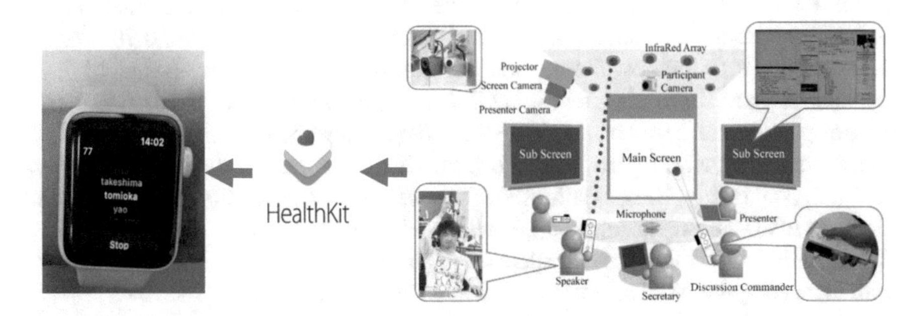

Fig. 1 Data collection system

4.2 Answers' Appropriateness Prediction Methodology

Feature Extraction

There were 18 HR and HRV features computed from all Q&A segments as well as the question and answer periods separately, which include the mean, standard deviation (std.), variance, and spectral entropy of each period. The Root Mean Square of Successive Difference (RMSSD) was also calculated from these three periods. The trends (upward or downward) of HR data in these three periods were also computed.

In order to extract meaningful text-based related cues as a predictor to identify the answer-appropriateness of Q&A segments in the discussion, an independent dataset from 41 lab seminar discussions recorded by the DM system for about 10 months during 2013 and 2014 was generated, with in total, 1,246 Q&A segments extracted as survey data samples. These Q&A segments' answer-appropriateness were also evaluated manually as the same way with the experimental dataset. Then TF-IDF (Term Frequency-Inverse Document Frequency) method [19] was adopted to extract 14 meaningful bi-grams from both classes as text-based features.

Supervised Classifiers and Validation

A set of supervised classifiers is built, including logistic regression, support vector machine, and random forest, based on single heart rate modality, text-based modality along with the combination of heart rate and text-based modality separately to predict the appropriateness of answers in Q&A. In addition, Recursive Features Elimination (RFE) [15] is used (only on the training dataset) to rank the features according to their importance to the different evaluation models, which used the RFE with the fivefold cross-validation (RFECV) to determine the best size of the feature subset. Leave-one-student-out cross-validation was performed to evaluate the performance of each detector to ensure generalization to new students.

4.3 Results

The first bar group in Fig. 2 shows the classification results of the classifiers built with HR modality, in which all of the classifiers yield AUC scores over 0.70. These remarkable results suggest the outstanding ability of HR modality in identifying students' answer appropriateness in a Q&A session. On the other hand, classifiers were also generated based on single text-based modality and the combination of HR and text modality, whose performance is shown in the second and the third bar groups in Fig. 2. Apparently, HR modality shows an overall stronger ability than text-based modality in classifying the appropriateness level of students' answers. Furthermore, HR modality can be seen to help add external information over text-based modality with increasing the AUC scores by an average 20% for all of the classifiers.

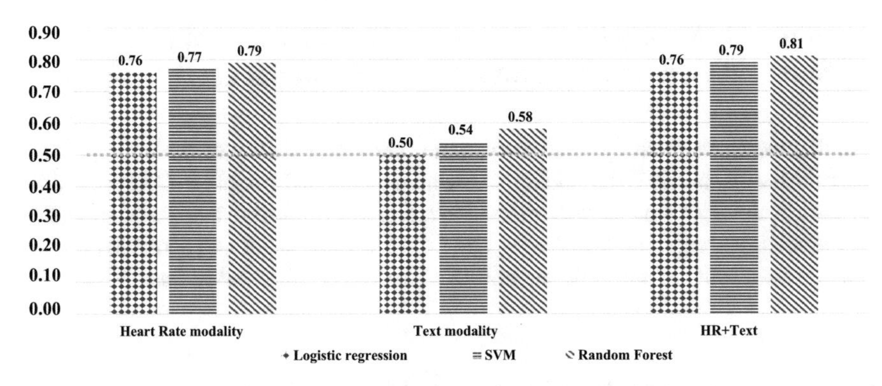

Fig. 2 AUC scores of each classifier with different modalities

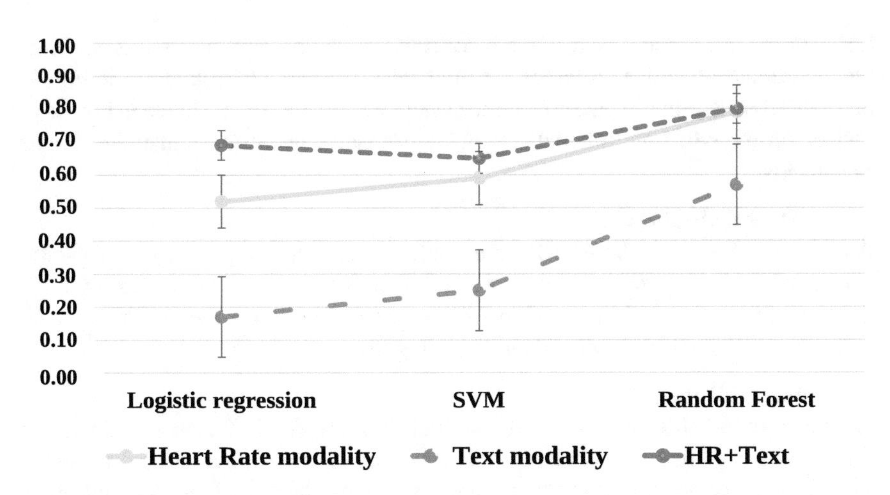

Fig. 3 Recall score regarding inappropriate answer detection

From an educational point of view, the identification ability of the classifier needs to be considered in effectively retrieving the inappropriate answers given by students, thereby to alert a teacher to provide timely adaptive coaching supports. Figure 3 shows the recall scores of classifiers based on different modalities. The solid line shows the recall score of classifiers based on single HR modality. The RF classifier achieved an AUC score with 0.80 which suggests a strong ability of HR modality in identifying students' inappropriate answers in a face-to-face Q&A session. Meanwhile, the text-based modality did a bad job in the task of inappropriate answers recognition. Furthermore, adding HR modality over text-based modality clearly improves the identification ability.

5 Study II: Recognition of Students' Multiple Mental States in Conversation Based on Multimodal Cues

The first exploitative study verified the better predictive ability of students' physiological cue, that is, the HR modality, in revealing students' trouble in Q&A sessions. Therefore, the HR modality was selected to replace the high-cost text-based modality as a prediction cue, in addition to two other modalities, facial and acoustic cues, to characterize students' multiple mental states in conversation-based discussions.

5.1 Multimodal Dataset Collection Methodology

Data for the multimodal dataset were collected on the basis of a group of students having discussion with a professor in a university's research lab, as shown in Fig. 4a and a multi-sensor-based data collection system was developed as shown in Fig. 4b to collect "in-the-wild" multimodal data involving facial, heart rate, and audio signals of students.

Visual Data Collection
As shown in Fig. 4(1), the ARKit framework was used to integrate the front-facing camera of the iPhone to detect the face and track the positions of the face with six degrees of freedom, and then a virtual mesh was overlaid over the face to simulate facial expressions in real time, as shown in Fig. 5

Physiological Data Collection
As shown in Fig. 4(3), the same with the first study, an Apple Watch paired with an iPhone was used to detect students' changes in heart rate for the entire discussion. The students were asked to wear the Apple Watch on their wrist, and synchronized with the iPhone.

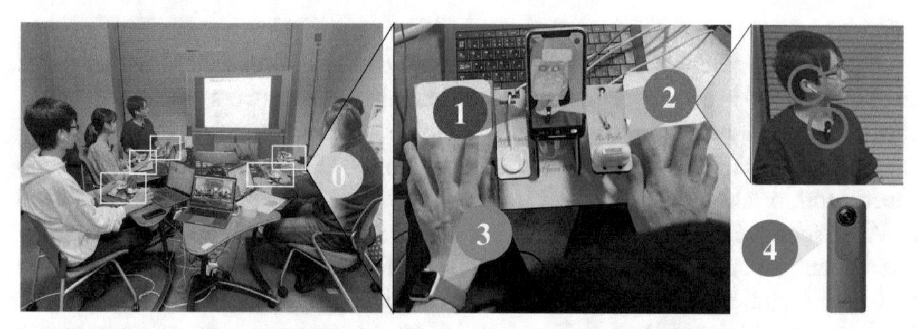

(a) Experimental scenario **(b) Multimodal data collection**

Fig. 4 **a** Small group face-to-face conversation-based coaching discussion. (0) Data collection system was placed on a desk in front of each participant. **b** Multi-sensor-based data collection system. (1) ARKit running on iPhone for face tracking. (2) AirPods for recording audio. (3) Apple Watch for detecting heart rate. (4) Ricoh THETA to record panorama video of the experiment in 360°

Fig. 5 Virtual mesh generated through face tracking results to simulate facial expressions

Audio Data Collection
AirPods were used to synchronously record the participants' audio data, as shown in Fig. 4(2). In the case that the audio data could not be recorded due to equipment failure or lack of power, students were required to wear pin microphones to record their audio data. Cloud Speech-to-Text was used to convert speech content into text content and then a manual text revision tool was designed to manually modify the text that was translated incorrectly while listening to a recording of the speaker's speech as well as add a period after each complete sentence so that entire speech statements could be divided into sentence units. These text data in sentence units were used as subtitles for the video in the annotation work explained in the next section.

360 degree's Panorama Video Recording
For the panorama video recording, as shown in Fig. 4(4), Ricoh THETA was set in the middle of the participants.

Observer Retrospective Annotation of Mental States
Two independent annotators were employed including one professor and one Ph.D. course student to annotate mental states of speaker-students by observing their facial expressions, upper body movements (since the participants were sitting, most of the body movements occurred on the upper body, such as raising hands and body leaning forward), and speech cues (speed, manner, tempo, and text content). A video-audio-based retrospective annotation tool was developed as shown in Fig. 6.

Multimodal Dataset
A total of 10 meetings were recorded, accumulating around 2,000 min worth of video-audio and physiological data with a mean length of approximately 500 min for each student. In the end, 1,772 successful observations of mental states were obtained which received consistent judgment from the two annotators, and used as the ground truth of the mental states of students in this study. Looking at the details of the data, concentration was the most common mental state observed by annotators (75.2%), followed by frustration (10.4%), confusion (9.6%), and boredom (4.8%).

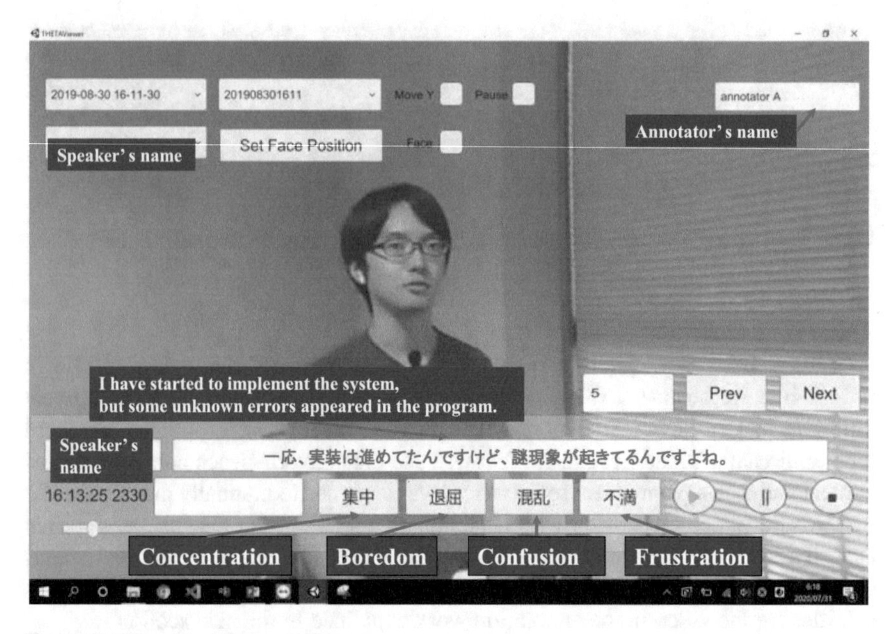

Fig. 6 Tool for annotating mental states

5.2 Methodology for Recognizing Multiple Mental States

Extracting Heart Rate Features
Two types of heart-rate-related features were extracted as follows:

- Aggregated heart-rate features: One of the methods was deriving a series of simple statistic features including the mean, standard deviation (std.), Root Mean Square Successive Difference (RMSSD), max, min, variance, slope, mean gradient, and spectral entropy for the entire segments.
- Sequential pattern heart-rate features: In the second study, rich feature representations that can describe the moment-by-moment dynamic changes in the HR value using Symbolic Aggregate Approximation (SAX) were also explored [24, 25], which was done in two steps. First, the Piecewise Aggregate Approximation (PAA) [5] algorithm was applied to the standardized raw sampled heart-rate time series $T = \{t_1 ... t_n\}$, with zero mean and unit variance, where T is the time of each speech video segment. The time series of length T seconds were divided into $w(w = 5)$ equal-length segments and represented the w-dimensional space with a real vector $\overline{T} = \{\overline{t}_1 ... \overline{t}_w\}$, where the ith element was computed with the following $\overline{T}$ following Eq. 1:

$$\overline{t}_i = \frac{w}{n} \sum_{j=\frac{n}{w}(j-1)+1}^{\frac{n}{w}i} t_j \tag{1}$$

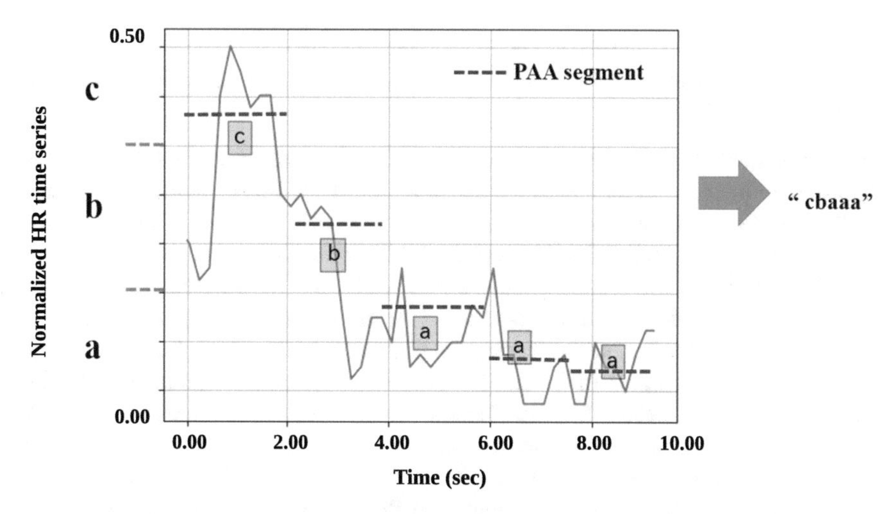

Fig. 7 Example of sequences generated from HR time series using SAX representation

Second, the PAA sequences of values were mapped into a finite list of symbols. The discretion threshold was chosen so that the distribution of symbols was approximately uniform. An alphabet of size 3 {a, b, c} was chosen to represent the PAA sequences to reflect the underlying dynamics of heart-rate transition among three levels, i.e., low, medium, and high. In Fig. 7, an example of the SAX representation "cbaaa" generated from a raw heart-rate time series is shown as a way of characterizing temporal dynamic patterns. Then, a feature representative method, "Bag-of-Words," where each word is a SAX pattern such as "cbaaa." was used. Altogether, there were 243 "words" of HR SAX patterns.

Extracting Acoustic Features

The INTERSPEECH 2009 Emotion Challenge feature set derived from openSMILE [10] was used, which contains 384 standard audio features that have been validated in terms of prediction ability regarding the task of recognizing emotion/mental states. These features are based on 16 base contours (MFCC 1–12, RMS energy, F0, zero crossing rate, and HNR) and their first derivatives (with 10 ms time windows). Features for a whole chunk were obtained by applying 12 functions [mean, standard deviation, kurtosis, skewness, minimum and maximum value, relative position, and range as well as two linear regression coefficients with their Mean Square Error (MSE)].

Extracting Facial Features

A series of dynamic facial features describing the movement patterns of the eye and mouth were extracted at an average frequency of 30 Hz. The first 300 frames (10 s) from each entire meeting video were used as a baseline in computing the features. Eye- and mouth-related dynamic events were measured for a given time window of

3 sec. and then several statistical features including mean, standard derivation (std.), max, min, range, and Root Mean Square (RMS) over the entire video segments were computed.

(1) Eye-Related Features: Coefficients describing the changes in the closure of the eyelids over the left and right eyes were used to detect eye-blink events. The average of the eyelids' movement coefficient of both eyes was measured when Pearson's r score was higher than 0.70. However, when a head rotation outside this range was detected, as it often happens in "in-the-wild" uncontrolled environments as in this study, only the movement coefficient of the visible eye was used. The raw eyelid movement coefficient time series was further denoised using a Savitzky-Golay filter [33] with a window of 15 frames to remove artifacts introduced when the device occasionally lost track of faces, leading to incorrect measurement. Then, peak detection [7] methods were applied to detect the local maximum (peak, eye-shut) and local minimum (valley, eye-opening). Eye blinks were detected by identifying a complete cycle from open (low coefficient) to close (high coefficient) and then back to open. Fake blinks were filtered by setting a threshold of 0.50 as the minimum peak coefficient since it may indicate eye squinting and a minimum between-peak duration of 0.40 s since an eye-blink cycle is around 0.40 to 0.60 s. The eye-blink frequency was estimated on the basis of the detected eye-blink events as one of the eye-related features. In addition, two other related features were derived to describe the sustained duration of eye-closure and eye-opening. Presumably, when a student's concentration level is heightened, the duration for which their eyes remain open may increase, while eyes closed for a long period of time may indicate that a student is squinting or is feeling bored.

(2) Mouth-Related Features: The "smiling" state is defined as when two corners of the mouth, respectively, appear in the first and the fourth quadrants with a minimum movement coefficient of 0.30, and when the movement of the corners of the mouth is detected in the second or the third quadrant with a maximum movement coefficient of −0.30, the state at this time is regraded as the "frowning" state. Then, the sustained durations of "smiling" and "frowning" were computed by measuring the position of both mouth corners in 2D space. In addition, the velocity and acceleration of mouth open-close movements were also measured along the vertical direction by computing lip movement coefficients. Besides those basic patterns of mouth movements, from the data itself, a state of interest is also found, that is, the corners of the mouth show a slanted line, which means that the movement of the corners of the mouth is detected in the first (or second) and the third (or fourth) quadrants. Presumably, when a student is confused or frustrated with the discussion content, the movements in the mouth area would appear diagonal. To filter out fake "diagonal lines" that occur due to actions made when speaking, a threshold is set for an effective slanted line, that is, when the slope of the line between the corners of the mouth falls between −0.57 and 0.57. Those patterns of mouth movements are shown in Fig. 8.

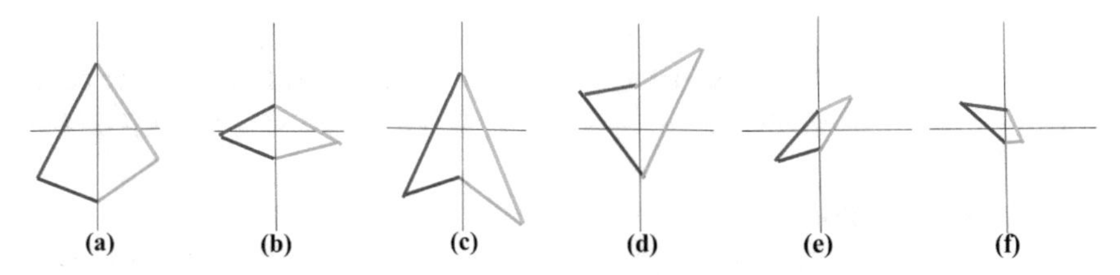

Fig. 8 Example of mouth movement patterns observed from one student when he conversed with the teacher: left and right corners of lips and middle points of upper and lower lips. **a** Mouth open, **b** mouth close, **c** frown, **d** smile, **e** the movement of the corners of the mouth is detected in the first and third quadrants, **f** the movement of the corners of the mouth is detected in the second and fourth quadrants

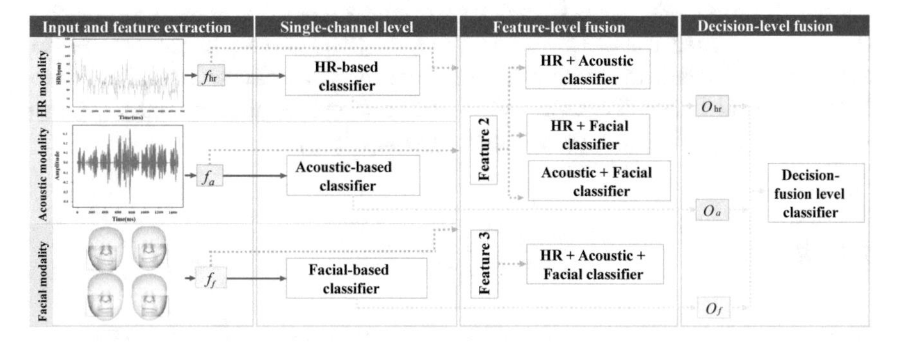

Fig. 9 Supervised learning mental state prediction classifiers with different multimodal fusion approaches

Supervised Classification of Students' Multiple Mental States

A line of supervised learning models with three different multimodal fusion methods was built as shown in Fig. 9. The feature vectors for each modality are denoted as f_{hr}, f_a, and f_f, which, respectively, represent a set of modalities of $M = \{heart\ rate, acoustic, and facial\}$. First, three baseline prediction models were separately built on the basis of individual channels: f_{hr}, f_a, and f_f. Second, four feature-level fusion prediction models were built in which the three modalities were combined together and a multi-label classifier was trained along with three other classifiers based on two modalities each time. For feature-level fusion prediction models, considering the different numbers of each modality's features, RELIEF-F was applied to extract the important features of each modality regarding the prediction tasks. Finally, decision-fusion-level classification models were also built, in which three single-channel-level classifiers were used as base classifiers to make classifications on the same test instances separately. Then, voting was performed on the prediction results (the probability of belonging to each category) which were denoted as O_{hr}, O_a, and O_f, respectively, and the result of the base classifier with the highest decision probability was selected as the final decision of each instance. The advantage of building decision-level fusion learning models is that, even in the case that some of the modality information was corrupted due to signal noise, was missing, or could not be captured due to occlusion or sensor artifacts, etc., we could still train the predictive learning models on the instances using a decision-level fusion approach.

Classifiers and Validation
The Synthetic Minority Oversampling Technique (SMOTE [6]) is adopted in the training data (not applied to the testing data) to control data imbalance in order to improve model fit. Then, a set of multi-class classifiers was built based on three kinds of supervised-learning machine learning models including Support Vector Machine (SVM), Random Forest (RF), and Multi-layer Perceptron network (MLP). Leave-one-student-out cross-validation was performed to evaluate the prediction performance of each classifier at the student level to ensure generalization to new students. The aggregate Area Under Curve (AUC) of the ROC curve (chance level = 0.5) scores is reported here to measure the overall performance of each classifier in identifying all mental state classes as well as the performance of each modality fusion method in recognizing each mental state class separately.

5.3 *Results*

Table 1 presents the aggregate AUC scores of each multi-class classifier in recognizing all mental state classes and the models that achieved the best overall performance (in bold) along with the feature numbers used from each modality when the best performance was yielded.

For the overall performance of single-channel classifiers, the RF classifiers (using 300 trees, along with balanced class weights and hyperparameter optimization using randomized search with 100 iterations) did a better job both in using the single HR modality and in using the single acoustic modality to recognize students' multiple mental states by demonstrating a better identification capability. Among them, the RF classifier based on the HR modality achieved a mean AUC of 0.704, which was

Table 1 Mean AUC scores of each classifier with different modality fusion approaches

Fusion approaches	SVM	RF	MLP	No. features
HR	0.673	**0.704**	0.690	10
Facial	0.651	0.716	**0.718**	15
Acoustic	0.694	**0.728**	0.721	20
HR + Acoustic	0.683	**0.759**	0.737	30
HR + Facial	0.680	**0.725**	0.724	25
Acoustic + Facial	0.685	**0.739**	0.731	35
HR + Acoustic + Facial	0.701	**0.763**	0.741	45
Decision-voting	0.687	0.733	**0.734**	⋆

slightly better than the MLP classifier with a mean AUC of 0.690. In addition, there were 10 top-ranked HR-related features that were used to generate the HR-modality-based classifiers where the best performance is achieved. Meanwhile, the RF classifier also showed outstanding performance in using acoustic cues to recognize all of the mental state classes, which yielded a mean AUC of 0.728, stronger than the SVM classifier with a mean AUC of 0.694 and better than the MLP classifier with a mean AUC score of 0.721. However, a difference was noticed for the facial-based single-channel classifiers, that is, the MLP model (seven layers, with active function of relu along with using cross-entropy as loss function) could learn the facial features better than the other two supervised learning models, with a mean AUC score of 0.718, a small advantage over the RF model, which had a mean AUC of 0.716, but stronger than the SVM model, which had a mean AUC of 0.651. Also noticed that fusing channels provided a significantly notable overall recognition performance improvements over each individual channel. Among them, fusing the HR and acoustic channels helped with improving the overall recognition performance by increasing the mean AUC scores by 5.5% over the individual HR channel and by 3.1% over the acoustic individual channel. In particular, combining the three modalities (AUC = 0.763) showed the best recognition ability over using only any single modality and any of the other combination methods in identifying the mental states of students.

The performance of the classifiers was also examined based on each individual channel and the classifiers based on the feature-fusion level in discriminating each mental state class as shown in Fig. 10. In addition, both of classifiers are based on the RF learning models that achieved the best performance as presented above. From the perspective of helping teachers augment their just-in-time decision-making capabilities, it is more interesting to effectively recognize students' states of confusion and frustration as much as possible. As shown in the second bar group of Fig. 10, fusing the HR and acoustic modalities (fourth bar, AUC = 0.695) was more effective than any of the other fusion approaches. In addition, fusing those two modalities helped with improving the prediction ability over only using the single HR modality or acoustic modality, with increasing the AUC score by 6.5 and 3.3%, respectively. Meanwhile, for identifying the state of frustration, the combination of three modali-

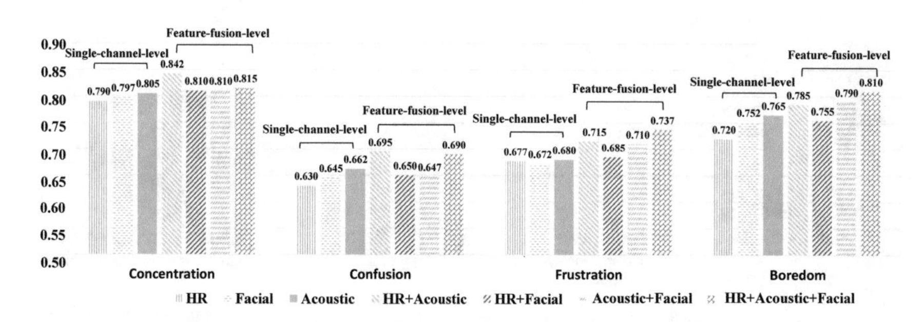

Fig. 10 Recognition performance for each mental state class using different multimodal fusion approaches

ties achieved the best recognition performance with an AUC of 0.737 as shown in the third bar group of Fig. 10. These results indicate that multimodal data can provide complementary information to each other, which results in augmenting the overall recognition ability in identifying the mental states of students.

6 Real-World Data Circulation to Educational Support

Real-World Educational Data Collection
For the target applications, applying the agent both for offline and online discussion learning environments involving multiple participants is expected, as well as for the activity of students learning through an online tutoring system, which have been a common self-learning method. However, for this case, the monitoring of learning situations is often not sufficient. The lack of external intervention may lead to high drop-off rates or other aspects of sub-optimal learning outcomes. Figures 11 and 12 present the application of the proposed system in real-world environments in both offline and online learning settings.

Data Analysis
Then, this information is input into the proposed mental state recognition models to detect the mental states of students. For the cases of multi-participant coaching activities, the proposed mental state detector can assist "busy" teachers not only to effectively monitor multiple students' mental states, but also to take advantage of

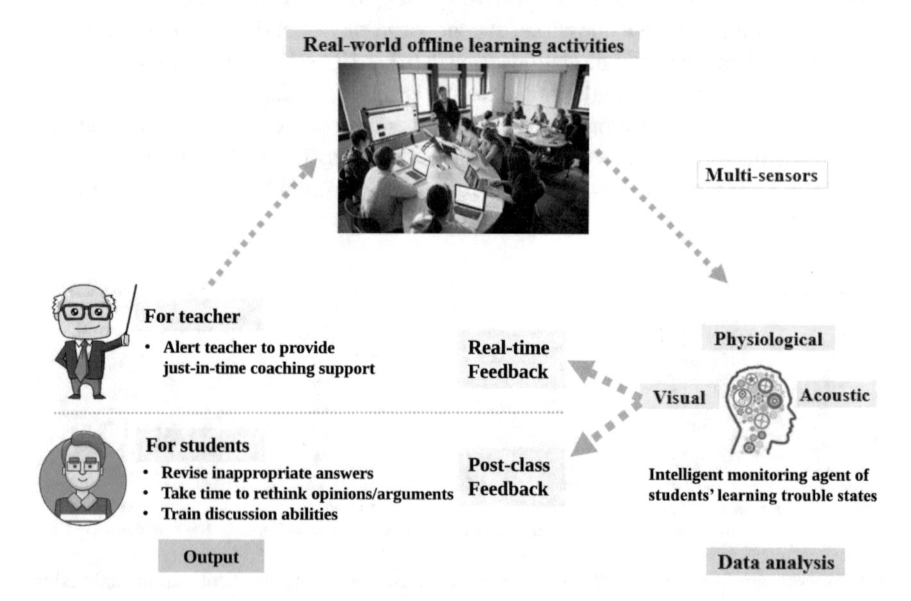

Fig. 11 Using the proposed system in a real-world environment of an offline discussion activity

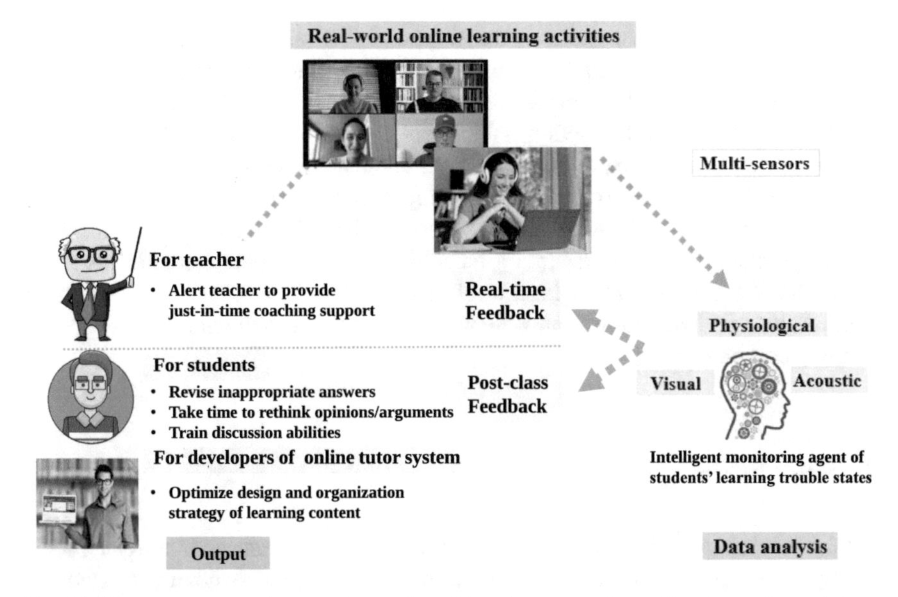

Fig. 12 Using the proposed system in a real-world environment of an online discussion activity

the replacement capabilities between different combinations of modalities to provide human teachers with alternate solutions for addressing the challenges with monitoring students such as when, in an online learning activity, the visual or audio modalities may not be available when non-speakers shut off their camera and mute their microphone, or in an offline learning activity, the facial modality is sometimes not available when students are wearing a mask. Similarly, the proposed system could also act as a virtual teacher with high practicality to support students learning through an online tutoring system.

Output

Furthermore, several feedback functions can be considered for help cultivating learning outcomes. To help a teacher make just-in-time decisions when providing instructions, designing a real-time feedback mechanism to alert teachers of the negative states of students during learning should be useful, such as when they are confused with opinions, or when they started feeling hopeless or frustrated with the learning content. This is because, if such negative states persist, the lack of external intervention could result in students losing interest and motivation in learning, which could hinder the smooth progress of learning and ultimately decrease learning outcomes.

In addition, it is not recommended to always give feedback to students in real time because this will interfere with their thinking process, in which case, those unresolved contents can be returned as post-class materials. For the purpose of training, it would be helpful to summarize and give feedback on students' unresolved content after discussion and to encourage them to spend more time reconsidering the appropriate answers or reorganizing and thinking about the opinions/comments that they did not understand during the discussion. Another type of post-class feedback that would be helpful to teachers is to help them discover which discussion content the students showed a high level of concentration for and where the students became bored and sleepy. This will assist teachers in effectively designing and arranging the lecture items or discussion strategies to increase the students' interest in learning as much as possible. Meanwhile, except for providing post-class feedback to students and teachers, it would be useful to design post-class feedback functions for online learning tutoring system developers regarding what kind of learning content students show clear concentration for with a high level of interest and for what kind of learning content students lose motivation. This will help system developers to optimize the learning components.

7 Conclusion

In this research, machine intelligence was leveraged to generate an intelligent monitoring agent to provide teachers with solutions to challenge with monitoring students regarding the recognition of multiple mental states including certainty level, concentration, confusion, frustration, and boredom in different educational environments.

To achieve these goals, an advanced multi-sensor-based data collection system was developed to accumulate and archive a "in-the-wild" massive multimodal dataset, for which the Apple Watch was used for real-time heart-rate data detection, ARKit framework was integrated with iPhone front-camera to track and collect facial motion signals, and AirPods with pin microphones were used to record the audio of discussions. A series of interpretable proxy features from visual, physiological, and audio modalities were derived separately to characterize multiple mental states. Then, a set of supervised learning SVM, RF, and MLP classifiers were trained separately using different multimodal fusion approaches including single-channel-level, feature-level, and decision-level fusion for recognizing students' multiple mental states.

From the results of this research, it is suggested that a student's physiological data can be used as an effective predictor in recognizing their learning troubles in discussion. In addition, this research has explored how to design multimodal analytics to augment the ability to recognize different mental states and found that fusing heart rate and acoustic modalities yields better recognition of the states of concentration (AUC = 0.842) and confusion (AUC = 0.695), while fusing three modalities yields the best performance in recognizing the states of frustration (AUC = 0.737) and boredom (AUC = 0.810). Furthermore, the results also explored the possibility of leveraging the advantages of the replacement capabilities between different modalities to provide human teachers with solutions for addressing the challenges with monitoring students in different real-world education environments.

References

1. Bosch N, D'mello SK, Ocumpaugh J, Baker RS, Shute V, (2016) Using video to automatically detect learner affect in computer-enabled classrooms. ACM Trans Inter Intel Syst 6(2):1–26
2. Bruner JS (2009) The process of education. Harvard University Press
3. Calvo RA, D'Mello SK (2011) New perspectives on affect and learning technologies, vol 3. Springer Science & Business Media
4. Castellano G, Kessous L, Caridakis G (2008) Emotion recognition through multiple modalities: face, body gesture, speech. In: Affect and emotion in human-computer interaction. Springer, pp 92–103
5. Chakrabarti K, Keogh E, Mehrotra S, Pazzani M (2002) Locally adaptive dimensionality reduction for indexing large time series databases. ACM Trans Database Syst 27(2):188–228
6. Chawla NV, Bowyer KW, Hall LO, Kegelmeyer WP (2002) Smote: synthetic minority oversampling technique. J Artif Intel Res 16:321–357
7. Du P, Kibbe WA, Lin SM (2006) Improved peak detection in mass spectrum by incorporating continuous wavelet transform-based pattern matching. Bioinformatics 22(17):2059–2065
8. D'Mello S, Graesser A (2012) Dynamics of affective states during complex learning. Learn Instr 22(2):145–157
9. D'Mello S, Craig S, Fike K, Graesser A (2009) Responding to learners' cognitive-affective states with supportive and shakeup dialogues. In: Proceeding of 2009 international conference on human-computer interaction, pp 595–604
10. Eyben F, Wöllmer M, Schuller B (2010) Opensmile: the munich versatile and fast opensource audio feature extractor. In: Proceedings of the 18th ACM international conference on multimedia, pp 1459–1462
11. Forbes-Riley K, Litman D (2011) When does disengagement correlate with learning in spoken dialog computer tutoring? In: Proceedings of 2011 international conference on artificial intelligence in education, pp 81–89
12. Gomes J, Yassine M, Worsley M, Blikstein P (2013) Analysing engineering expertise of high school students using eye tracking and multimodal learning analytics. In: Proceedings of the 6th international conference on educational data mining 2013, pp 375–377
13. Grafsgaard J, Wiggins JB, Boyer KE, Wiebe EN, Lester J (2013a) Automatically recognizing facial expression: predicting engagement and frustration. In: Proceedings of the 6th international conference on educational data mining 2013, pp 43–50
14. Grafsgaard JF, Wiggins JB, Boyer KE, Wiebe EN, Lester JC (2013b) Embodied affect in tutorial dialogue: Student gesture and posture. In: Proceeding of 2013 international conference on artificial intelligence in education, pp 1–10
15. Guyon I, Weston J, Barnhill S, Vapnik V (2002) Gene selection for cancer classification using support vector machines. Mach Learn 46(1–3):389–422
16. Hellhammer J, Schubert M (2012) The physiological response to trier social stress test relates to subjective measures of stress during but not before or after the test. Psychoneuroendocrinology 37(1):119–124
17. Hoque ME, McDuff DJ, Picard RW (2012) Exploring temporal patterns in classifying frustrated and delighted smiles. IEEE Trans Affect Comput 3(3):323–334
18. Hussain MS, AlZoubi O, Calvo RA, D'Mello SK (2011) Affect detection from multichannel physiology during learning sessions with autotutor. In: Proceeding of 2011 international conference on artificial intelligence in education, pp 131–138
19. Jones KS (1972) A statistical interpretation of term specificity and its application in retrieval. J Docum 28(1):11–20
20. Koedinger KR, Aleven V (2007) Exploring the assistance dilemma in experiments with cognitive tutors. Educat Psychol Rev 19(3):239–264
21. Kovanović V, Joksimović S, Waters Z, Gašević D, Kitto K, Hatala M, Siemens G (2016) Towards automated content analysis of discussion transcripts: a cognitive presence case. In: Proceedings of the 6th international conference on learning analytics & knowledge, pp 15–24

22. Landis JR, Koch GG (1977) The measurement of observer agreement for categorical data. Biometrics 33:159–174
23. Lepper MR, Woolverton M, Mumme DL, Gurtner J (1993) Motivational techniques of expert human tutors: lessons for the design of computer-based tutors. Comput Cognit Tools 1993:75–105
24. Lin J, Keogh E, Lonardi S, Chiu B (2003) A symbolic representation of time series, with implications for streaming algorithms. In: Proceedings of the 8th ACM SIGMOD workshop on research issues in data mining and knowledge discovery, pp 2–11
25. Lin J, Keogh E, Wei L, Lonardi S (2007) Experiencing sax: a novel symbolic representation of time series. Data Mining Knowl Discov 15(2):107–144
26. Monkaresi H, Bosch N, Calvo RA, D'Mello SK (2016) Automated detection of engagement using video-based estimation of facial expressions and heart rate. IEEE Trans Affect Comput 8(1):15–28
27. Mukherjee S, Yadav R, Yung I, Zajdel DP, Oken BS (2011) Sensitivity to mental effort and test-retest reliability of heart rate variability measures in healthy seniors. Clinical Neurophys 122(10):2059–2066
28. Nagao K, Kaji K, Yamamoto D, Tomobe H (2004) Discussion mining: annotation-based knowledge discovery from real world activities. In: Proceeding of 2004 Pacific-rim conference on multimedia, pp 522–531
29. Peng S, Chen L, Gao C, Tong RJ (2020) Predicting students' attention level with interpretable facial and head dynamic features in an online tutoring system (student abstract). In: Proceedings of the 34th international conference on association for the advancement of artificial intelligence, pp 13895–13896
30. Pereira T, Almeida PR, Cunha JP, Aguiar A (2017) Heart rate variability metrics for fine-grained stress level assessment. Comput Methods Progr Biomed 148:71–80
31. Reilly JM, Schneider B (2019) Predicting the quality of collaborative problem solving through linguistic analysis of discourse. Proceeding of the 12th international conference on educational data mining society, pp 149–157
32. Rodrigo MMT, Baker RS, Agapito J, Nabos J, Repalam MC, Reyes SS, San Pedro MOC (2012) The effects of an interactive software agent on student affective dynamics while using; an intelligent tutoring system. IEEE Trans Affect Comput 3(2):224–236
33. Savitzky A, Golay MJ (1964) Smoothing and differentiation of data by simplified least squares procedures. Anal Chem 36(8):1627–1639

Towards Practically Applicable Quantitative Information Flow Analysis

Bao Trung Chu

Abstract An adaptive information security solution is always demanded, especially in this fast changing digital world. This adaptivity would not likely be achieved by a single method, but rather by multiple approaches that are well-organized in a system. Based on the philosophy of real-world data circulation, the model of an adaptive system like that is introduced in the beginning of this chapter. Attempts to speed up the static analysis of quantitative information flow and analyze the run-time information leakage, which help software verifying and monitoring in an adaptive solution, respectively, follows later. A discussion on building a practical adaptive system in the future, then, sums up the chapter.

1 Data Circulation in Information Security

Figure 1 represents the life cycle of a software starting from *Programming* in which a product is implemented. The development of a new software is the loop between *Programming* and *Verification*. Following this loop, the quality of a software will be improved incrementally both in terms of usability and security. Once software passes predefined requirements in *Verification*, the product is released for *Running* in real-world applications. While being used, the released software is monitored for possible failures which could be either functional bugs or security vulnerabilities. When *Monitoring* detects a failure, the software will be sent back to the very first block for root cause analysis and fixing. Again, software goes through the loop, is released, monitored, and so forth. Unless being replaced by a brand new software using a novel technology or there is no need anymore, a software will travel through this life cycle infinitely.

In the life cycle of a software, information security is considered at *Verification* and *Monitoring*. While the former is static, the latter requires solutions during running time. Figure 2 illustrates two cycles, each of which includes three phases in a data circulation system: Acquisition, Analysis, and Implementation. The cycle on the left

B. T. Chu (✉)
Super Zero Lab, Tokyo, Japan
e-mail: trungchubao@superzero-lab.com

K. Takeda et al. (eds.), *Frontiers of Digital Transformation*,
https://doi.org/10.1007/978-981-15-1358-9_5

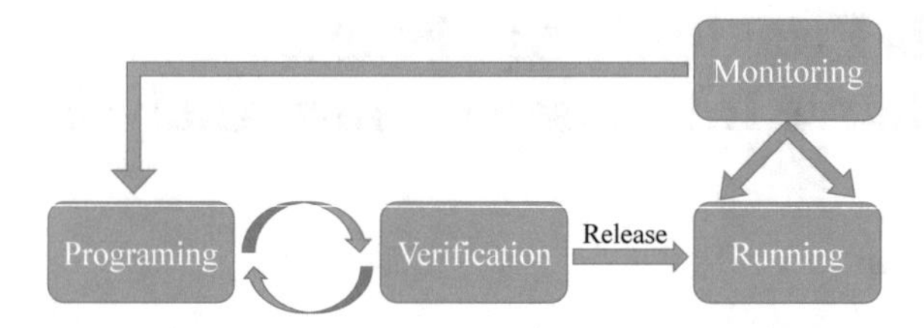

Fig. 1 Software life cycle

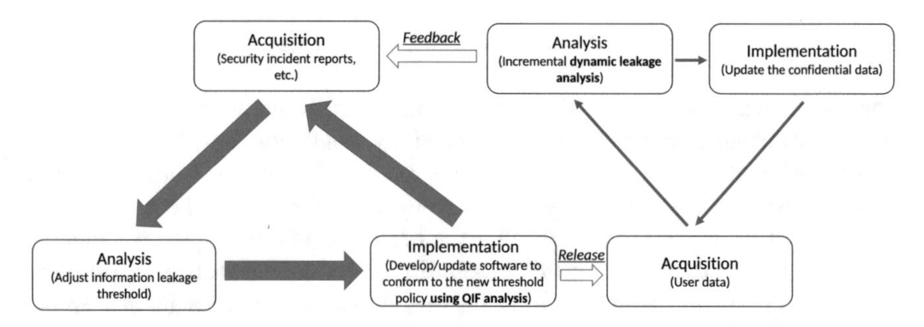

Fig. 2 Data circulation in information security

corresponds to the information security enforcing process for *Verification*. The other on the right describes the process for *Monitoring*. The two cycles communicate back and forth through Release and Feedback which again creates one big cycle. In each loop, there are chances where decisions need to be made for updating the software so that it can adapt well to change in the real-world. In the implementation phase of the loop on the left, it is decided if the software satisfies the security policy or not. While in the implementation phase of the loop on the right, it is decided if confidential data need to be updated. In both of the cases, a quantitative criterion for making decisions is desirable. Thus Quantitative Information Flow (QIF) just goes in as a good fit for the adaptive system.

2 Introduction to Quantitative Information Flow

In this section, the definitions of QIF based on Shannon entropy and Min entropy as well as model counting-based calculation methods are introduced. The content is mainly based on [13]. In the remaining parts of this section we use the following notations. Let P be the program under analysis of QIF, $\mathcal{S}$ be the finite set of input values and $\mathcal{O}$ be the finite set of output values of P. Without loss of generality, we assume that P has one secret input variable $s \in \mathcal{S}$ and one publicly observable output variable $o \in \mathcal{O}$. Random variables S and O represent the events in which s takes a value from $\mathcal{S}$ and o takes a value from $\mathcal{O}$, respectively.

2.1 Entropy-Based QIF

Given random variables X and Y ranging over finite sets $\mathcal{X}$ and $\mathcal{Y}$, respectively, Shannon entropy over X is defined as

$$H(X) = \sum_{x \in \mathcal{X}} p(X = x) \log \frac{1}{p(X = x)}, \tag{1}$$

and the conditional entropy of X, provided that an event associated with the random variable Y took place, is

$$H(X|Y) = \sum_{y \in \mathcal{Y}} p(Y = y) \sum_{x \in \mathcal{X}} p(X = x|Y = y) \log \frac{1}{p(X = x|Y = y)}, \tag{2}$$

where $p(X = x)$ denotes the probability that X has value x, and $p(X = x|Y = y)$ denotes the probability that X has value x when Y has value y. Informally, QIF is defined by the equation:

$$\text{QIF} = \text{Initial Uncertainty} - \text{Remaining Uncertainty}. \tag{3}$$

In (3), initial uncertainty and remaining uncertainty are the degrees of surprise on the value of s **before** and **after** executing P, respectively. Therefore we have the Shannon entropy-based QIF of P defined as

$$\text{QIF}^{\text{Shannon}}(P) = H(S) - H(S|O). \tag{4}$$

For deterministic programs, the information leakage is simplified as $H(O)$.

Example 1 Let us consider the following example borrowed from [13].

$$o := s \;\&\; 0\text{x}037,$$

where s is an unsigned uniformly distributed 32-bit integer, and hence $0 \leq s \leq 2^{32}$. We have $H(S) = 2^{32}\frac{1}{2^{32}}\log 2^{32} = 32$ bits, $H(S|O) = 2^5\frac{1}{2^5}(2^{27}\frac{1}{2^{27}}\log 2^{27}) = 27$ bits. By (4), $\text{QIF}^{\text{Shannon}}(P) = 5$ bits.

Smith [13] proposes a different notion called Min entropy-based QIF. Firstly, he defines the vulnerability over a random variable X as

$$V(X) = \max_{x \in \mathcal{X}} p(X = x), \tag{5}$$

and the conditional vulnerability of X given Y is

$$V(X|Y) = \sum_{y \in \mathcal{Y}} p(Y = y) \max_{x \in \mathcal{X}} p(X = x|Y = y). \tag{6}$$

Min entropy over X is defined as

$$H_\infty(X) = \log \frac{1}{V(X)}, \tag{7}$$

and the conditional Min entropy of X given Y is

$$H_\infty(X|Y) = \log \frac{1}{V(X|Y)}. \tag{8}$$

Thus by (3), Min entropy-based QIF of P is defined as the following:

$$\mathrm{QIF}^{\min}(P) = H_\infty(S) - H_\infty(S|O). \tag{9}$$

Especially when P is deterministic and $\mathcal{S}$ is uniformly distributed, (9) is simplified to

$$\mathrm{QIF}^{\min}(P) = \log|\mathcal{S}| - \log\frac{|\mathcal{S}|}{|\mathcal{O}|} = \log|\mathcal{O}|, \tag{10}$$

where $|\mathcal{S}|$ and $|\mathcal{O}|$ denote the cardinalities of $\mathcal{S}$ and $\mathcal{O}$, respectively.

2.2 *Model Counting-Based Calculation of QIF*

This subsection is dedicated to the introduction of two popular methods to calculate QIF by model counting. As required by model counters, let us assume the program under analysis P be deterministic and terminating, and the input value set $\mathcal{S}$ is uniformly distributed. This assumption technically can be removed by using the result in [2].

A. Satisfiability (SAT)-based Counting

When $\mathcal{S}$ is uniformly distributed, we have $\mathrm{QIF}^{\min}(P) = \log|\mathcal{O}|$ and $\mathrm{QIF}^{\mathrm{Shannon}}(P) = H(O)$. It suffices to count the distinct output values to calculate the former, while the distribution on $\mathcal{O}$ is needed to calculate the latter. Let $\mathcal{O} = \{o_1, o_2, ...o_n\}$ such that $o_i \neq o_j\ \forall i \neq j$. For each $o_i \in \mathcal{O}$, let $C_{o_i} = \{s' \in \mathcal{S} \mid P(s') = o_i\}$. Because P is deterministic, we have $C_{o_i} \cap C_{o_j} = \emptyset\ (\forall i \neq j)$. C_{o_i} is called the preimage set of o_i by program P. The probability distribution on $\mathcal{O}$ is $p(o = o_1) = \frac{|C_{o_1}|}{|\mathcal{S}|}$, $p(o = o_2) = \frac{|C_{o_2}|}{|\mathcal{S}|}$, ..., $p(o = o_n) = \frac{|C_{o_n}|}{|\mathcal{S}|}$. Thus,

$$\mathrm{QIF}^{\mathrm{Shannon}}(P) = H(O) = \sum_{o_i \in \mathcal{O}} \frac{|C_{o_i}|}{|\mathcal{S}|} \log \frac{|\mathcal{S}|}{|C_{o_i}|}. \tag{11}$$

Let $\langle P \rangle$ denote the propositional formula in CNF of P, $(b_1 b_2 ... b_k)$ and $(r_1 r_2 ... r_t)$ be the propositional representation of s and o, respectively. Without loss of generality, we assume that $\langle P \rangle$ has one input and one output. Basically, besides the propositional variables corresponding to the input and the output, $\langle P \rangle$ includes many not-of-interest variables. It is necessary, therefore, to project models onto the set of Boolean variables that represent the input and the output, to count the distinct output values. Let Δ be the set of projection variables. Projecting a model m onto Δ is to remove from the model all the truth values of propositional variables that are not in Δ. Algorithm 1 illustrates the counting method. Model($\langle P \rangle$) (lines 2 and 6) invokes a SAT solver and it returns a model of $\langle P \rangle$. In the right hand side of line 5, $\Delta \not\equiv m|_\Delta$ expresses the constraint that all the variables in Δ are different from their counterparts in the projection of model m on Δ. This constraint assures that all models in M are distinct.

Algorithm 1 SATBasedProjectionCount($\langle P \rangle$, Δ) [8]

```
1: M ← ∅
2: m ← Model(⟨P⟩)
3: while m ≠ ⊥ do
4:     M ← M ∩ m|Δ
5:     ⟨P⟩ ← ⟨P⟩ ∧(Δ ≢ m|Δ)
6:     m ← Model(⟨P⟩)
7: return M
```

B. Satisfiability Modulo Theory (SMT)-based Counting

An SMT-based counting method is proposed by Phan et al. [11, 12]. Assume $\mathcal{V}$, $\mathcal{F}$, and $\mathcal{P}$ be countable sets of variables, function symbols, and predicate symbols, respectively.

Definition 1 (*SMT* [11]) Given a theory or a combination of theories $\mathcal{T}$ and a Σ-formula φ, the problem SMT is the problem of deciding $\mathcal{T}$-satisfiability of φ.

Definition 2 (*#SMT* [11]) Given a theory or a combination of theories $\mathcal{T}$ and a Σ-formula φ, the Model Counting Modulo Theories problem (#SMT) is the problem of finding all models M of $\mathcal{T}$ w.r.t. a set of Boolean variables V_I such that φ is $\mathcal{T}$-satisfiable in M.

Because there are theories $\mathcal{T}$ that allow infinite range of variables, the set V_I of *important* variables is added to assure the number of models w.r.t. V_I to be finite. The basic difference between #SAT and #SMT is that the former works on Boolean variables after transforming original constraints into this basic unit of propositional logic, while the latter directly works on original constraints.

Besides φ and V_I, V_R is another argument that specifies *relevant* variables, namely, the assignments of the variables of interest when a model is found w.r.t. V_I. V_R is designed for the automated test generation, but not directly related to calculating QIF. In Algorithm 2: N and Ψ are the number of models and the enumeration of the models; Assert(φ) in line 2 and Assert(block) in line 10 are to set the Σ-formula φ in the problem #SMT, by line 10, φ is reassigned to $\varphi\wedge$ block; Check() in line 3 is to verify if φ is $\mathcal{T}$-satisfiable or not; Model(φ) in line 5 is to retrieve one model in

Algorithm 2 All-BC(φ, V_I, V_R) [11]

1: $N \leftarrow 0$; $\Psi \leftarrow \varepsilon$;
2: Assert(φ);
3: **while** Check() = SAT **do**
4: $N \leftarrow N + 1$;
5: $m \leftarrow$ Model(φ);
6: $m_{ir} \leftarrow$ Filter(m, V_I, V_R);
7: $\Psi \leftarrow \Psi \cup \{m_{ir}\}$;
8: block $\leftarrow$ FALSE;
9: **for all** $p_i \in V_I$ **do** block $\leftarrow$ block $\vee$ ($p_i \neq$ Eval(p_i));
10: Assert(block);
11: **return** N, Ψ;

case φ is $\mathcal{T}$-satisfiable; Filter(m, V_I, V_R) in line 6 is to extract from model m the counterparts of V_I and V_R; Lines 8 and 9 are used to add the blocking clause to the Σ-formula so that the same model will not be retrieved again. The idea is similar to line 5 in Algorithm 1.

3 Attempt to Speed Up Software Verification

Calculating QIF, an important step in software verification, relies hugely on model counting which is costly especially for complex data structures like strings. Here a novel counting method for string constraints based on Context-Free Grammars (CFG) is proposed. Figure 3 is an example of a string constraint. The dot "." represents a string concatenation. `ASSERT(T_4)` indicates a verification if there is any assignment to the string variables that makes `T_4` true. The only solution for this example is `var_0xINPUT_2` = `"Hello=Joe"`.

The lengths of strings to count must be bounded to assure the number of solutions is finite. As illustrated in Fig. 4, combining a string counter with a transformer like

Fig. 3 A sample string constraint

```
PCTEMP_LHS_1 := T1_1.T2_1;
T2_1 == "=Online";
T1_1 == var_0xINPUT_2;
T_2 := T1_4.T2_4;
T2_4 == "Now";
T1_4 == PCTEMP_LHS_1;
T3 := T_2!="Hello=Joe=OnlineNow";
T_4 := !T_3;
ASSERT(T_4);
```

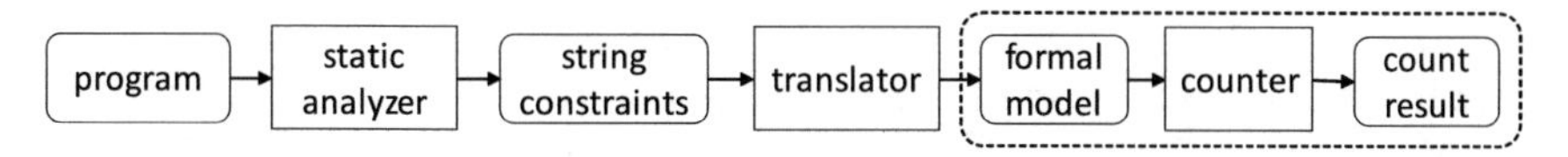

Fig. 4 String counting-based QIF calculation

Java String Analyzer[1] and PHP String Analyzer,[2] that can extract string constraints on the output of the program under analysis, is another solution to calculate QIF.

3.1 Context-Free Grammar-Based String Counting

A *polynomial* $S \in \mathbb{R}\langle\langle A\rangle\rangle$ is a formal series such that $\mathrm{supp}(S)$ is finite. The class of all polynomials over A with coefficients in $\mathbb{R}$ is denoted as $\mathbb{R}\langle A\rangle$. We fix an alphabet A and a semiring $\mathbb{R}$ and let Y be a countable set of variables disjoint with A.

Definition 3 (*algebraic system*) An ($\mathbb{R}$-) *algebraic system* is a finite set of equations of the form

$$y_i = p_i \quad (1 \leq i \leq n)$$

where $y_i \in Y$ and $p_i \in \mathbb{R}\langle A \cup Y\rangle$. The system is called *proper* if $(p_i, \varepsilon) = (p_i, y_j) = 0$ for all i, j $(1 \leq i, j \leq n)$. □

For a vector $\mathbf{v}$, let $\mathbf{v}^T$ denote the transpose of $\mathbf{v}$. Formally, let $R = (R_1, \ldots, R_n)^T \in \mathbb{R}\langle\langle A\rangle\rangle^{n\times 1}$ be an n-column vector of formal series and define the morphism $h_R : (A \cup Y)^* \to \mathbb{R}\langle\langle A\rangle\rangle$ by $h_R(y_i) = R_i$ $(1 \leq i \leq n)$ and $h_R(a) = a$ for $a \in A$. We furthermore extend h_R to a polynomial $p \in \mathbb{R}\langle A \cup Y\rangle$ by

$$h_R(p) = \sum_{w\in(A\cup Y)^*} (p, w)h_R(w).$$

R is a *solution* of the algebraic system $y_i = p_i$ $(1 \leq i \leq n)$ if $R_i = h_R(p_i)$ for all i $(1 \leq i \leq n)$.

We say $S \in \mathbb{R}\langle\langle A\rangle\rangle$ is *quasi-regular* if $(S, \varepsilon) = 0$. It is well-known that every proper $\mathbb{R}$-algebraic system $y_i = p_i$ $(1 \leq i \leq n)$ has a unique solution $R = (R_1, \ldots, R_n)^T$ such that every component R_i $(1 \leq i \leq n)$ is quasi-regular. R is called the *strong* solution of the system. The strong solution of a proper algebraic system $y_i = p_i$ $(1 \leq i \leq n)$ can be given by the limit $R = \lim_{j\to\infty} R^j$ of the sequence R^j defined by $R^0 = (0, \ldots, 0)^T$ and $R^{j+1} = (h_{R^j}(p_1), \ldots, h_{R^j}(p_n))^T$ for $j \geq 0$.

Example 2 For an $\mathbb{N}$-algebraic system $y = ayb + ab$, $R^0 = 0$, $R^1 = ab$, $R^2 = ab + a^2b^2, \ldots$, and $R = \sum_{n\geq 1} a^n b^n$. □

[1] https://www.brics.dk/JSA/.

[2] https://sv.c.titech.ac.jp/minamide/phpsa/.

Definition 4 (*algebraic series*) A formal series $S \in \mathbb{R}\langle\langle A\rangle\rangle$ is $\mathbb{R}$-*algebraic* if $S = (S, \varepsilon)\varepsilon + S'$ where S' is some component of the strong solution of a proper $\mathbb{R}$-algebraic system. The class of $\mathbb{R}$-algebraic series over A is denoted by $\mathbb{R}^{\mathtt{alg}}\langle\langle A\rangle\rangle$. □

A Context-Free Grammar (CFG) is a tuple (N, T, P, S) where N and T are finite sets of nonterminal and terminal symbols, respectively, P is a finite set of (production) rules in the form $A \to \gamma$ where $A \in N$ and $\gamma \in (N \cup T)^*$, and $S \in N$ is the start symbol. The derivation relation $\overset{*}{\Rightarrow}_G$ is defined in the usual way. For $A \in N$, let $L_G(A) = \{w \in T^* \mid A \overset{*}{\Rightarrow}_G w\}$. The context-free language generated by G is $L(G) = L_G(S)$.

Proposition 1 ([7, 9]) *A language L is context-free if and only if L is $\mathbb{N}$-algebraic if and only if L is $\mathbb{B}$-algebraic.*

Theorem 1 *Let $\alpha : y_i = p_i$ $(1 \le i \le n)$ be an $\mathbb{R}$-algebraic system in Chomsky normal form and let $R = (R_1, \dots, R_n)^T$ be the strong solution of α. Then, for each i $(1 \le i \le n)$ and $d \in \mathbb{N}$,*

$$CC(R_i, d) = \sum_{y_j y_m \in supp(p_i)\cap Y^2} (p_i, y_j y_m) \sum_{k=1}^{d-1} CC(R_j, k) CC(R_m, d-k)$$
$$+ \sum_{a \in supp(p_i)\cap A} (p_i, a)\Delta[d = 1]. \tag{12}$$

Moreover, $\mathrm{CC}(R_i, d)$ can be computed in $O(\xi d^2)$ time where $\xi = n \max_{1\le i\le n} |p_i|$ [3]. We can extend the above algorithm to apply to an arbitrary algebraic series.[3]

Example 3 Consider the following algebraic system:

$$y_1 = y_2 y_3 + y_4 y_5, \quad y_2 = a y_2 + a, \quad y_3 = b y_3 c + bc,$$
$$y_4 = a y_4 b + ab, \quad y_5 = c y_5 + c.$$

Let $R = (R_1, \dots, R_5)^T$ be the strong solution of the system. $\mathrm{CC}(R_i, d)$ $(1 \le i \le 5,$ $1 \le d \le 5)$ can be computed by Theorem 1 as shown below.

	R_1	R_2	R_3	R_4	R_5
1	0	1	0	0	1
2	0	1	1	1	1
3	2	1	0	0	1
4	2	1	1	1	1
5	4	1	0	0	1

[3]We can keep the time complexity $O(\xi d^2)$ through this extension by revising the algorithm in such a way that an equation of length three or more is virtually divided into equations of length two on the fly.

Also,

$$\mathrm{CC}(R_1, 6) = \sum_{k=1}^{5}(\mathrm{CC}(R_2, k)\mathrm{CC}(R_3, 6-k) + \mathrm{CC}(R_4, k)\mathrm{CC}(R_5, 6-k)) = 4.$$

$R_1 = \sum_{i \neq j, i \geq 1, j \geq 1}(a^i b^j c^j + a^i b^i c^j) + \sum_{i \geq 1} 2a^i b^i c^i$ and $\mathrm{supp}(R_1) = \{a^i b^j c^j \mid i, j \geq 1\} \cup \{a^i b^i c^j \mid i, j \geq 1\}$ is an (inherently) ambiguous context-free language. □

3.2 Experimental Evaluation

To evaluate the efficiency and applicability of the algorithm proposed in Section 4, the following experiments are conducted. Based on Theorem 1, a prototype was implemented for solving the following string counting problem for CFGs.

Input: CFG $G = (N, T, P, S)$, $A \in N$ and $d \in \mathbb{N}$.
Output: $\Sigma_{d' \leq d} |L_A(G)_{d'}|$, the number of strings derivable from A in G and of length not greater than d.

The tool was implemented in C++ and the experiments were conducted in the environment: Intel Core i7-6500U CPU@2.5GHz x 4, 8GB RAM, Ubuntu 16.10 64bits. GMP library was utilized to deal with big numbers.

Here, the experimental results conducted by the prototype are reported and compared to the state-of-the-art string counter ABC on Kaluza benchmark [1, 10]. Kaluza bechmark consists of two parts, *small* and *big*. The former includes 17,554 files and the latter does 1,342 files. This experiment was conducted on Kaluza small.

The ABC source code and benchmark were cloned from: https://github.com/vlab-cs-ucsb/ABC.[4] Both ABC and the proposed tool were run at length bounds 1, 2, and 3 in order to manually confirm the counts. There was no big difference about running time between the tools noticed. The comparison results (Table 1) were consistent through different length bounds (1, 2, and 3) in which ABC and the proposed tool gave the same results on 17,183 files. The other 371 files were manually investigated and were found that they can be classified into 21 groups such that all constraints in a same group are very similar. Among those 21 groups, the proposed tool gave correct answers for 18 groups, more precise than ABC for 2 groups, and less precise than ABC for 1 group. Overall, the experimental result shows that the proposed counting algorithm is potentially applicable to a qualified benchmark and it is promising to extend the algorithm.

[4]There were several branches without mentioning a stable release in the repository. So, the master branch as of 2017/12/27 was cloned.

Table 1 Proposed prototype versus ABC on Kaluza *small* at length bounds 1, 2, and 3

Proposed ≡ ABC	Proposed ≢ ABC			
17,183 files	*Proposed right*	*Proposed better*	*ABC better*	*ABC right*
	227 files	143 files	0 files	1 file
	(18 groups)	(2 groups)	(0 groups)	(1 group)
	371 files			
Total = 17,544 files				

4 Quantitative Approach for Software Monitoring

In Example 4, assume $source$ to be a positive integer, then there are 16 possible values of $output$, from 8 to 23. While an observable value between 9 and 23 reveals *everything* about the secret variable, a value of 8 gives almost nothing. It is crucial to differentiate risky execution paths from safe ones. Here, a couple of notions to model this amount of information, so-called dynamic leakage, are proposed.

Example 4 if source $<$ 16 then output $\leftarrow 8 +$ source
else output $\leftarrow 8$

4.1 New Notions

Let P be a program with a secret input variable S and an observable output variable O. For notational convenience, the names of program variables are identified by the corresponding random variables. We assume that a program always terminates. Also let $\mathcal{S}$ be the set of values of S and $\mathcal{O}$ be the set of values of O. For $s \in \mathcal{S}$ and $o \in \mathcal{O}$, let $p_{SO}(s, o)$, $p_{O|S}(o|s)$, $p_{S|O}(s|o)$, $p_S(s)$, $p_O(o)$ denote the joint probability of $s \in \mathcal{S}$ and $o \in \mathcal{O}$, the conditional probability of $o \in \mathcal{O}$ given $s \in \mathcal{S}$ (the likelihood), the conditional probability of $s \in \mathcal{S}$ given $o \in \mathcal{O}$ (the posterior probability), the marginal probability of $s \in \mathcal{S}$ (the prior probability) and the marginal probability of $o \in \mathcal{O}$, respectively. We often omit the subscripts as $p(s, o)$ and $p(o|s)$ if they are clear from the context. By definition,

$$p(s, o) = p(s|o)p(o) = p(o|s)p(s), \tag{13}$$

$$p(o) = \sum_{s \in \mathcal{S}} p(s, o), \tag{14}$$

$$p(s) = \sum_{o \in \mathcal{O}} p(s, o). \tag{15}$$

We assume that (the source code of) P and the prior probability $p(s)$ $(s \in \mathcal{S})$ are known to an attacker. For $o \in \mathcal{O}$, let $\mathrm{pre}_P(o) = \{s \in \mathcal{S} \mid p(s|o) > 0\}$, which is called the preimage of o (by the program P).

The first proposed notion is the self-information of the secret inputs consistent with an observed output $o \in \mathcal{O}$ (see the upper part of Fig. 5).

$$\mathrm{QIF}_1(o) = -\log\left(\sum_{s' \in \mathrm{pre}_P(o)} p(s')\right). \tag{16}$$

The second notion is the self-information of the joint events $s' \in \mathcal{S}$ and an observed output $o \in \mathcal{O}$ (see the lower part of Fig. 5).

$$\mathrm{QIF}_2(o) = -\log\left(\sum_{s' \in \mathcal{S}} p(s', o)\right) = -\log p(s, o) + \log p(s|o). \tag{17}$$

Theorem 2 *If a program P is deterministic, for every $o \in \mathcal{O}$ and $s \in \mathcal{S}$,*

$$\mathrm{QIF}^{\mathrm{belief}}(s, o) = \mathrm{QIF}_1(o) = \mathrm{QIF}_2(o) = -\log p(o),$$

and if input values are uniformly distributed, for every $o \in \mathcal{O}$,

$$\mathrm{QIF}_1(o) = \log \frac{|\mathcal{S}|}{|\mathrm{pre}_P(o)|}$$

□

In Example 4, we assume: both *source* and *output* are 8-bit numbers in the range of 0 and 255, and source is uniformly distributed. Then, because the program is deterministic QIF_1 coincides with QIF_2. We have $\mathrm{QIF}_1(\mathrm{output} = 8) = -\log \frac{241}{256} = 0.087$ bits, $\mathrm{QIF}_1(\mathrm{output} = o) = -\log \frac{1}{256} = 8$ bits for every o between 9 and 23.

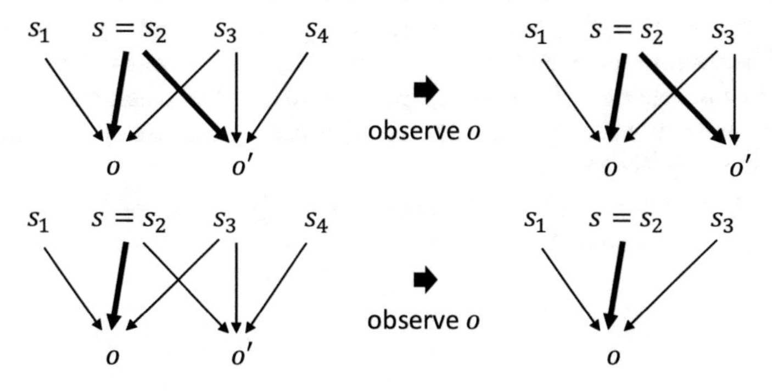

Fig. 5 QIF_1 (the upper) and QIF_2 (the lower)

4.2 Program Model

A program P has the following syntax:

$$P ::= \text{in } \mathbf{S};\ \text{out } \mathbf{O};\ \text{local } \mathbf{Z};\ c \mid P;\ P$$

where $\mathbf{S}$, $\mathbf{O}$, and $\mathbf{Z}$ are sequences of variables which are disjoint from one another. We first define $\text{In}(P)$, $\text{Out}(P)$, and $\text{Local}(P)$ for a program P as follows:

- If $P = \text{in } \mathbf{S};\ \text{out } \mathbf{O};\ \text{local } \mathbf{Z};\ c$, we define $\text{In}(P) = \{V \mid V \text{ appears in } \mathbf{S}\}$, $\text{Out}(P) = \{V \mid V \text{ appears in } \mathbf{O}\}$ and $\text{Local}(P) = \{V \mid V \text{ appears in } \mathbf{Z}\}$. In this case, we say P is a simple program. We require that no variable in $\text{In}(P)$ appears on the left-hand side of an assignment command in P, i.e., any input variable is not updated.
- If $P = P_1;\ P_2$, we define $\text{In}(P) = \text{In}(P_1)$, $\text{Out}(P) = \text{Out}(P_2)$ where we require that $\text{In}(P_2) = \text{Out}(P_1)$ holds. We also define $\text{Local}(P) = \text{Local}(P_1) \cup \text{Local}(P_2) \cup \text{Out}(P_1)$.

A program P is also written as $P(S, O)$ where S and O are enumerations of $\text{In}(P)$ and $\text{Out}(P)$, respectively. A program $P_1;\ P_2$ represents the sequential composition of P_1 and P_2. Note that the semantics of $P_1;\ P_2$ is defined in the same way as that of the concatenation of commands $c_1;\ c_2$ except that the input and output variables are not always shared by P_1 and P_2 in the sequential composition.

4.3 Model Counting-Based Calculation of Dynamic Leakage

Although computing dynamic leakage is proved to be a very hard problem [4], it is possible to reduce computations of QIF_1 and QIF_2 to model counting in some reasonable assumptions. Let us consider what is needed to compute based on their definitions (16) and (17).

For calculating QIF_1 for a given output value o, it suffices (1) to enumerate input values s' that satisfy $p(s'|o) > 0$ (i.e., possible to produce o), and (2) to sum the prior probabilities over the enumerated input values s'. (2) can be computed from the prior probability distribution of input values, which is reasonable to assume. When input values are uniformly distributed, only step (1) is needed because QIF_1 is simplified to $\log \frac{|S|}{|\text{pre}_P(o)|}$ by Theorem 2.

As for QIF_2, when programs are *deterministic*, $\text{QIF}_1 = \text{QIF}_2$. We would leave computing QIF_2 for probabilistic programs as future work.

4.4 Compositional Approach to Speed Up Calculating Dynamic Leakage

Aiming at speeding up the model counting-based calculation of dynamic leakage, a "divide and conquer" fashioned method is proposed here. The dynamic leakage of the original program will be calculated from the leakage of constituted sub-programs, sequentially or in parallel or a combination of both.

Sequential Composition

This section proposes a method of computing both exact and approximated dynamic leakage by using sequential composition. For making the idea behind the proposed method understandable, the programs under analysis are assumed to be deterministic with uniformly distributed input, so that the problem of quantifying dynamic leakage is reduced to model counting.

A. Exact Calculation
For a program $P(S, O)$, an input value $s \in \mathcal{S}$ and a subset $\mathcal{S}'$ of input values, let

$$\text{post}_P(s) = \{o \mid p(o|s) > 0\},$$
$$\text{post}_P(\mathcal{S}') = \bigcup_{s \in \mathcal{S}'} \text{post}_P(s).$$

If P is deterministic and $\text{post}_P(s) = \{o\}$, we write $\text{post}_P(s) = o$.

Let $P = P_1; P_2$ be a program. We assume that $\text{In}(P_1)$, $\text{Out}(P_1)$, $\text{In}(P_2)$, $\text{Out}(P_2)$ are all singleton sets for simplicity. This assumption does not lose generality; for example, if $\text{In}(P_1)$ contains more than one variables, we instead introduce a new input variable that stores the tuple consisting of a value of each variable in $\text{In}(P_1)$. Let $\text{In}(P) = \text{In}(P_1) = \{S\}$, $\text{Out}(P_1) = \text{In}(P_2) = \{T\}$, $\text{Out}(P) = \text{Out}(P_2) = \{O\}$, and let $\mathcal{S}, \mathcal{T}, \mathcal{O}$ be the corresponding sets of values, respectively. For a given $o \in \mathcal{O}$, $\text{pre}_P(o)$ and $p(o)$ can be represented in terms of those of P_1 and P_2 as follows.

$$\text{pre}_P(o) = \bigcup_{t \in (\text{pre}_{P_2}(o) \cap \text{post}_{P_1}(\mathcal{S}))} \text{pre}_{P_1}(t), \tag{18}$$

$$p(o) = \sum_{s \in \mathcal{S}, t \in \mathcal{T}} p(s) p_1(t|s) p_2(o|t). \tag{19}$$

If $p(s)$ is given, we can compute (18) by enumerating $\text{pre}_{P_1}(t)$ for $t \in (\text{pre}_{P_2}(o) \cap \text{post}_{P_1}(\mathcal{S}))$ and also for (19). This approach can easily be generalized to the sequential composition of more than two programs, in which the enumeration is proceeded in a Breadth-First-Search fashion. However, in this approach, search space will often explode rapidly and lose the advantage of composition. Therefore, we come up with an approximation, which is explained in the next subsection, as an alternative.

B. Approximation

Algorithm 3 LowerBound($P_1, \cdots, P_n, o$, timeout)

```
1: Pre[2..n] ← empty
2: Stack ← empty
3: level ← n
4: acc_count ← 0
5: Push(Stack, o)
6: Pre[n] ← EnumeratePre(P_n, o)
7: while not Stack.empty and execution_time < timeout do
8:    if level = 1 then
9:       acc_count ← acc_count+ CntPre(P_1, Stack.top)
10:      level ← level + 1
11:      Pop(Stack)
12:   else
13:      v ← PickNotSelected(Pre[level])
14:      if v = AllSelected then
15:         level ← level + 1
16:         Pop(Stack)
17:      else
18:         Push(Stack, v)
19:         level ← level − 1
20:         if level > 1 then
21:            Pre[level] ← EnumeratePre(P_level, v)
22: return acc_count
```

Lower bound:
To infer a lower bound of $|\mathrm{pre}_P(o)|$, we leverage Depth-First-Search (DFS) with a predefined timeout such that the algorithm will stop when the execution time exceeds the timeout and output the current result as the lower bound. The method is illustrated in Algorithm 3. For a program $P = P_1; P_2; \cdots; P_n$, an observable output o of the last sub-program P_n and a predetermined $timeout$, Algorithm 3 derives a lower bound of $|\mathrm{pre}_P(o)|$ by those n sub-programs. In Algorithm 3, CntPre*(Q, o)* counts $|\mathrm{pre}_Q(o)|$, PickNotSelected*(Pre[i])* selects an element of $Pre[i]$ that has not been traversed yet or returns *AllSelected* if there is no such element, and EnumeratePre*(P_i, v)* lists all elements in $\mathrm{pre}_{P_i}(v)$. $Pre[i]$ stores $\mathrm{pre}_{P_i}(o_i)$ for some o_i. For P_1, it is not necessary to store its preimage because we need only the size.

Upper bound:
For an upper bound of $|\mathrm{pre}_P(o)|$ we use Max#SAT problem [6], which is defined as follows.

Definition 5 Given a propositional formula $\varphi(X, Y, Z)$ over sets of variables X, Y, and Z, the Max#SAT problem is to determine $\max_X \#Y.\exists Z.\varphi(X, Y, Z)$.

If we consider a program Q, In(Q), Out(Q) and Local(Q) as φ, Y, X, and Z, respectively, then, the solution X to the Max#SAT problem can be interpreted as the output value, which has the biggest size of its preimage set. In other words, $\max_X \#Y.\exists Z.\varphi(X, Y, Z)$ is an upper bound of the size of pre_Q over all feasible outputs. Therefore, the product of those upper bounds of $|\mathrm{pre}_{P_i}|$ over all i $(1 \le i \le n)$ is obviously an upper bound of $|\mathrm{pre}_P|$. Algorithm 4 computes this upper bound where CntPre(P_n, o) returns the size of the preimage of o by P_n. Notice that, to avoid enumerating the preimages, which costs much computation time, we count only $|\mathrm{pre}_{P_n}(o)|$. For $i = 1, ...n-1$, we compute MaxCount(P_i) as an upper bound for pre_{P_i}, regardless of the corresponding output value.

Algorithm 4 UpperBound($P_1, \cdots, P_n, o$)

```
1: Result ← CntPre(P_n, o)
2: for i ← 1 to n do
3:     Result ← Result*MaxCount(P_i)
4: return Result
```

Value Domain Decomposition

Another effective method for computing the dynamic leakage in a compositional way is to decompose the sets of input values and output values into several subsets, compute the leakage for the sub-programs restricted to those subsets, and compose the results to obtain the leakage of the whole program.

Let $P(S, O)$ be a program. Assume that the sets of input values and output values, $\mathcal{S}$ and $\mathcal{O}$, are decomposed into mutually disjoint subsets as

$$\mathcal{S} = \mathcal{S}_1 \uplus \cdots \uplus \mathcal{S}_k,$$
$$\mathcal{O} = \mathcal{O}_1 \uplus \cdots \uplus \mathcal{O}_l.$$

For $1 \le i \le k$ and $1 \le j \le l$, let P_{ij} be the program obtained from P by restricting the set of input values to $\mathcal{S}_i$ and the set of output values to $\mathcal{O}_j$ where if the output value o of P for an input value $s \in \mathcal{S}_i$ does not belong to $\mathcal{O}_j$, the output value of P_{ij} for input s is undefined. In practice, this disjoint decomposition can be done simply by augmenting the program under analysis with appropriate constraints on input and output.

By definition, for a given $o \in \mathcal{O}_j$,

$$\mathrm{pre}_P(o) = \bigcup_{1 \le i \le k} \mathrm{pre}_{P_{ij}}(o). \tag{20}$$

By Theorem 2, if P is deterministic and the prior probability of S is uniformly distributed, what we have to compute is $|\mathrm{pre}_P(o)|$, which can be obtained by summing up each $|\mathrm{pre}_{P_{ij}}(o)|$ by (20):

$$|\text{pre}_P(o)| = \sum_{1 \le i \le k} |\text{pre}_{P_{i,j}}(o)|.$$

Calculating for probabilistic programs with arbitrary input distribution is a future work.

Experimental Result

This section investigates two questions: (1) How well does parallel computing based on the value domain decomposition improve the performance? and (2) How well does approximation in the sequential composition work in terms of precision and speed? Here, those questions are examined through examples: *Grade protocol* is for question (1), and *Population count* is for (2). The benchmarks and prototype are public.[5] The experiments were conducted on Intel(R) Xeon(R) CPU ES-1620 v3 @ 3.5GHz x 8 (4 cores x 2 threads), 32GB RAM, CentOS Linux 7. For parallel computation, OpenMP [5] library was used. A tool for Algorithm 3 and Algorithm 4, as well as the exact count in sequential compositions was implemented in Java.

A. Grade protocol
This benchmark is taken from [11]. The benchmark with *4 students* and *5 grades* was used, and all variables were of 16 bits. We suppose the observed output (the sum of students' grades) to be **1**, and hence the number of models is 4. The original program was manually decomposed into 4, 8, and 32 sub-programs based on the value domain decomposition: The set of output values was divided into 2 and the set of input values was divided into 2, 4, or 16 disjoint subsets.

In Table 2, n: number of sub-programs; t: number of threads specified using OpenMP. Note that $n = 1$ means non-decomposition, $t = 1$ means a sequential execution and the number of physical CPUs is 8. From the results, we can infer that parallel computing based on the value domain decomposition improves the performance at most 25 times, from 0.25s to 0.01s.

B. Population count
population_count is the 16-bit version of the benchmark of the same name given in [11]. In this experiment, the original program is decomposed into three sub-programs in such a way that each sub-program performs one-bit operation of the original. For model counting, we let the output value by 7 (so the number of models is 11440) for *population_count*.

Table 3 presents the execution times, exact count, lower bounds, and upper bound for model counting. The execution times for the lower bounds are predetermined timeouts, which were designed to be 1/2, 1/5, and 1/10 of the time needed by the exact count. From the results, we learn that (a) while over approximation speeds up the calculation almost 30 times, from 3.50s to 0.13s, the upper bound is more than 500 times bigger than the exact count; and (b) under approximation halves the

[5]https://bitbucket.org/trungchubao-nu/dla-composition/src/master/dla-composition/.

Table 2 Value domain decomposition based model counting time

	$n = 32$	$n = 8$	$n = 4$	$n = 1$
$t = 32$	0.05 s	–	–	–
$t = 16$	0.05 s	–	–	–
$t = 8$	0.05 s	0.01 s	–	–
$t = 4$	0.07 s	0.01 s	0.01 s	–
$t = 2$	0.12 s	0.02 s	0.02 s	–
$t = 1$	0.18 s	0.04 s	0.03 s	0.25 s

Table 3 Model counting: execution time and the changing of precision

	Execution time	Count
Exact count	3.50 s	11,440
Lower bound	1.75 s	4,712
	0.70 s	1,314
	0.35 s	52
	0.09 s	0
Upper bound	0.13 s	5,898,240

execution time, from 3.50s to 1.75s, while giving 4, 712 as the lower bound which is almost half of the exact count. Also noted that dynamic leakage is just the logarithm of counts, so the quantitative imprecision will be much smaller.

5 Conclusion

We are witnessing the transformation from analog to digital, real-world to virtual world, in an extremely fast pace that might never have happened before. We have used mobile phones, then smartphones, and now Internet of Things. We watched black & white TVs, then color TVs, and now virtual reality is coming closer. Internet becomes essential to our daily life. Many of us live two lives, real one and virtual one. Because of the trend, information security becomes much more relevant to us. On the other hand, real-world data, which includes all the data traveling on the internet, affects hugely the security. A fixed security policy will face difficulties in this rapidly changing world. While data circulation provides adaptability, QIF analysis offers flexibility to an **adaptive information security solution**. This mechanism might be a part of future cyber-security solutions.

References

1. Aydin A, Bang L, Bultan T (2015) Automata-based model counting for string constraints. In: Proceedings 27th international conference on computer aided verification, pp 255–272. LNCS, volume: 9206
2. Backes M, Berg M, Köpf B (2011) Non-uniform distributions in quantitative information flow. In: Proceedings 6th ACM symposium on information, computer and communications security, pp 367–375
3. Chu BT, Hashimoto K, Seki H (2018) Counting algorithms for recognizable and algebraic series. IEICE Trans Inform Syst E101-D(6):1479–1490
4. Chu BT, Hashimoto K, Seki H (2019) Quantifying dynamic leakage: complexity analysis and model counting-based calculation. IEICE Trans Inform Syst E102-D(10):1952–1965
5. Dagum L, Menon R (1998) OpenMP: an industry-standard API for shared-memory programming. IEEE Comput Sci Eng 5(1):46–55
6. Droste M, Kuich W, Vogler H (2009) Handbook of weighted automata. Springer
7. Fremont DJ, Rabe and MN, Seshia SA (2017) Maximum model counting. In: Proceedings 31st AAAI conference on artificial intelligence, pp 3885–3892
8. Klebanov V, Manthey N, Muise C (2013) SAT-based analysis and quantification of information flow in programs. In: Proceedings 10th international conference on quantitative evaluation of systems, pp 177–192
9. Kuich W, Salomma A (1986) Semirings, automata, languages. Springer
10. Luu L, Shinde S, Saxena P, Demsky B (2014) A model counter for constraints over unbounded strings. In: Proceedings 35th ACM SIGPLAN conference on programming language design and implementation, pp 565–576
11. Phan QS, Malacaria P (2015) All-solution satisfiability modulo theories: applications, algorithms and benchmarks. In: Proceedings 10th International conference on availability, reliability and security, pp 100–109
12. Phan QS (2015) Model counting modulo theories. In: Doctoral dissertation, Queen Mary University of London
13. Smith G (2009) On the foundations of quantitative information flow. In: Proceedings 12th internation conference on foundations of software science and computational structures, pp 288–302. LNCS, volume 5504

Frontiers in Mechanical Data Domain

Research on High-Performance High-Precision Elliptical Vibration Cutting

Hongjin Jung

Abstract Recent researches/developments about high-performance and high-precision elliptical vibration cutting and related topics about the Real-World Data Circulations are introduced in this chapter. The elliptical vibration cutting technology has been successfully utilized in the industry for the finishing of various precision parts and precision micro-machining of dies and molds, which are made of difficult-to-cut materials. However, applications of the elliptical vibration cutting have been limited to high-cost ultra-precision machining because of its high manufacturing cost and limited cutting conditions and usages. To overcome the problems and expand its applications, research on high-performance and high-precision elliptical vibration cutting is carried out in this work. Furthermore, the applied Real-World Data Circulation, i.e., data acquisition, analysis, and implementation, in this work is described as well.

1 Overview of Research on High-Performance High-Precision Elliptical Vibration Cutting

To realize the high-performance and high-precision ultrasonic elliptical vibration, the author carried out researches as follows.

1.1 Development of a High-Power Ultrasonic Elliptical Vibration Cutting Device

To expand the application of elliptical vibration cutting technology to low-cost mirror surface finishing, the development of a high-power elliptical vibration device whose

H. Jung (✉)
Department of Ultra-precision Machines & Systems, Korean Institute of Machinery & Materials, 156, Gajeongbuk-ro, Yuseong-gu, Daejeon 305-343, South Korea
e-mail: hj_jung@kimm.re.kr

K. Takeda et al. (eds.), *Frontiers of Digital Transformation*,
https://doi.org/10.1007/978-981-15-1358-9_6

available depth of cut is more than several hundreds of micrometers is required. The newly developed device should be designed considering two conflicting features at the same time. At first, it should be made larger compared to the conventional devices to make it sustain at larger cutting loads by adopting large-sized piezoelectric actuators which can load more electrical charge. Meanwhile, its size should be an acceptable size considering the practical usages. Therefore, the half-wavelength mode of axial vibration and dedicated support design are adopted. Then, to verify the performance of the high-power elliptical vibration cutting, mirror surface finishing of hardened steel is carried out at different cutting conditions. The experimental results such as cutting forces, surface roughness, and tool wear are evaluated. As a result, it is clarified that the newly developed vibration cutting device can increase the material removing rate and can be utilized in the ordinary manufacturing environments which leads to low manufacturing cost [1].

1.2 Clarification and Suppression of Chatter in Elliptical Vibration Cutting

As the depth or width of cut of elliptical vibration cutting is small, there are no undesirable vibration problems or the chatter. However, as the depth or width of cut gets larger, there will be a case where chatter occurs and cannot be neglected for further applications. The vibration is small like a micrometer but should be suppressed for obtaining mirror surfaces and avoid wear and breakage of extremely sharp diamond tools. Therefore, the generation mechanism of chatter is investigated by analyzing the finished surfaces and the undesirable vibrations superimposed on the elliptical vibration, and it is described regarding to the microprocess of the elliptical vibration cutting process, i.e. ploughing process and material removing process. Based on the generation mechanism of the chatter, a simple suppression method is also proposed and its validity is verified experimentally [2].

1.3 High-Precision Cutting of Titanium Alloy Ti-6Al-4 V

Ultrasonic elliptical vibration cutting of titanium alloy Ti-6Al-4 V is investigated in this research. Because products made of Ti-6Al-4 V alloy are usually designed for possessing low-rigidity structures or good-quality cut surfaces, machining requirements such as low cutting forces and slow rate of tool wear need to be fulfilled for the realization of their precision machining. Therefore, the ultrasonic elliptical vibration cutting is applied as a novel machining method for those products. Machinability of Ti-6Al-4 V alloy by the ultrasonic elliptical vibration cutting with cemented carbide tools is examined to figure out suitable cutting conditions for precision machining of Ti-6Al-4 V alloy. During the machining of the titanium alloy, forced vibration caused by the segmented chip generation, which is a well-known problem in ordinary non-vibration cutting, occurs. Therefore, the characteristics of the forced vibration due to the chip segmentation are investigated. Through the experiments, it is found that the frequency and magnitude of the forced vibration are related to the average uncut chip thickness and cutting width. Based on the clarified characteristics of forced vibration due to the chip segmentation, a practical strategy to suppress the forced vibration due to chip segmentation is experimentally examined and verified. As a result, it is possible to obtain a precision cut surface of titanium alloy with low cutting forces by applying the elliptical vibration cutting [3].

1.4 Process Monitoring of Elliptical Vibration Cutting

The importance of cutting process monitoring has increased under a recent trend toward manufacturing automation and the unmanned operation of a plant. Therefore, a new method for monitoring the elliptical vibration cutting process is proposed by utilizing internal data in the ultrasonic elliptical vibration device without external sensors such as a dynamometer and displacement sensors. The internal data used here is the change of excitation frequency, i.e., the resonant frequency of the device, voltages applied to the piezoelectric actuators composing the device, and electric currents flowing through the actuators. These internal data change automatically in the elliptical vibration control system to keep a constant elliptical vibration against the change of the cutting process. To estimate the cutting process from the internal data, vibration model of the elliptical vibration cutting process is established, and the modeling helps to analyze the internal data appropriately. Correlativity between the process and the internal data is described by using a proposed model of ultrasonic elliptical vibration cutting and verified by several experiments, i.e., planing and mirror surface finishing of hardened die steel carried out with single-crystalline diamond tools. As a result, it is proved that it is possible to estimate the elliptical vibration cutting process parameters, e.g., tool wear and machining load, which are crucial for stable cutting in such precision machining [4].

As a researcher of the Graduate Program for Real-World Data Circulation Leaders, the author tried to study how RWDC can contribute to the advanced cutting process. For example, in the thesis, the elliptical vibration tool controller's internal data is utilized to monitor the cutting process indirectly. The elliptical vibration tool controller is used only to stabilize the elliptical vibration. Still, a new value of the internal data is discovered by connecting, collecting, and analyzing the data based on the knowledge of the cutting process. The more specific and concrete conception for RWDC in the cutting process is demonstrated by introducing various related researches, and the novelty and contribution of the author's research thesis concerning RWDC are described in the following section.

2 Real-World Data Circulation in This Work

Real-World Data Circulation (RWDC) is a new academic field established by the Graduate Program for Real-World Data Circulation Leaders, Program for Leading Graduate Schools of Nagoya University. RWDC encompasses the acquisition, analysis, and implementation of real-world data. Moreover, it emphasizes the interactive integration of these three procedures to explore unknown mechanisms and create new real-world values. Digital data is acquired mainly through the observation or measurement of various phenomena in the real world. Those data should be analyzed with a practical and comprehensive manner to figure out and predict the mechanism of real-world phenomena. Moreover, those results should be implemented in the real world to solve the real-world problem or to improve the existing phenomena. A new approach with a perspective of RWDC will assist in understanding the real-world problem clearly which is intricately connected all over the society and for creating a new value that overcomes the stagnant general society circles such as humanities, economics, and manufacturing industries.

The author's primary concern is how RWDC will be activated to improve the manufacturing industry's circumstance, especially related to the cutting process. The ultrasonic elliptical vibration cutting, which is the main topic of this work, can also be considered an excellent example of RWDC. There is a real-world problem that the ordinary cutting of ferrous materials with a diamond tool is impossible because of the rapid tool wear. The elliptical vibration cutting method is implemented and it resolved the problem. Moreover, the real-world data related to the cutting process, such as cutting force, quality of cut surface, and tool wear, are acquired through the experiments. It is analyzed to figure out the mechanism of the ultrasonic elliptical vibration cutting process. As a result, it is possible to optimize the cutting conditions and vibration conditions and successfully implement it in the dies and molds industry after the feasibility of the process is confirmed.

Recently, the advanced technologies (e.g., networked machines, big data, and information analytics) are being introduced into the manufacturing environment. These integrations between information technologies and manufacturing technologies are encouraging the advent of the fourth industrial revolution, called Industry 4.0 [5]. The new era's industrial production will be highly flexible in production volume and customization, extensive integration between customers, companies, and suppliers, and above all sustainable [6]. Industry 4.0 describes production-oriented Cyber-Physical Systems (CPS) to understand the CPS which is essential for implementing the advanced manufacturing environment. The CPS refers to a new generation of systems with integrated computational and physical capabilities that can interact with humans through many new modalities [7, 8]. For example, the cutting process in a manufacturing site is a real-world physical system. Models of cutting processes that are proposed by researchers to reproduce the physical system as identical as possible computationally are representative examples of the cyber system. Data acquisition and analysis plays an important role in connecting and integrating the physical system and cyber systems. Digital simulation of the cutting process from

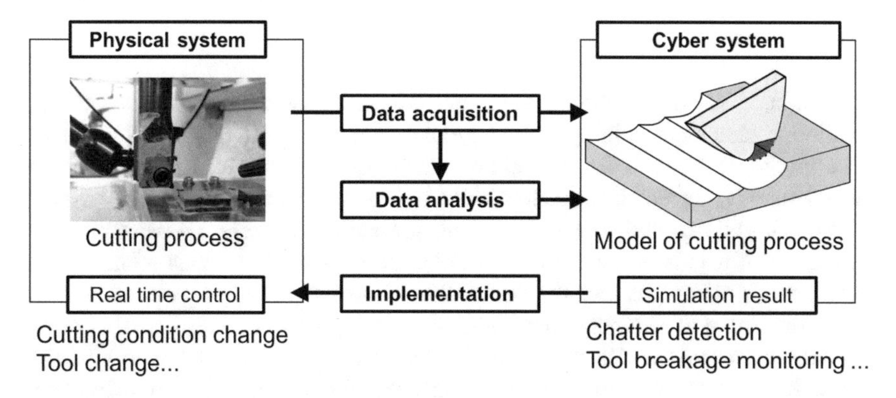

Fig. 1 Example of CPS in cutting process

the cyber systems should be integrated into real-world physical systems. Then the results can be implemented in the cutting process in real time (Fig. 1).

Meanwhile, to implement the CPS to the manufacturing site and utilize it practically to improve productivity, more efficient and low-cost data acquisition methods and advanced data analysis strategies should be developed and investigated. The promising fundamental techniques consisting of the CPS are cloud, IoT (Internet of Things), and big data [9]. There are lots of CNC (Computerized Numeric Control) machines, tools, products, and workers in the manufacturing site, and all of these should be connected through the cloud and IoT. This data acquisition state is called the "smart connection level" of 5C architecture for realizing CPS [10], which is the collection of all the data in the manufacturing site. The next level is the "data-to-information conversion level", which extracts the critical data or correlation among the manufacturing site's components for improving the cutting process. At this stage, advanced data analysis technologies such as machine learning are required. However, without the comprehensive knowledge and understanding of the cutting process itself, it is not efficient and reaches the limit for stepping over this level.

2.1 Real-World Data in Cutting Process and Its Acquisition

In the work, many kinds of data are used to figure out the mechanism of the cutting process. Data are acquired during the machining experiments or after the experiments. Then, the data are analyzed by various methods such as calculating the mean value, transforming the time-domain signals to frequency components, measuring the lengths from the photograph, etc. Through these kinds of processes, it is possible to understand the various phenomena in the cutting mechanism and improve the cutting process.

Cutting Force

The cutting force is essential to figure out the cutting process because the cutting conditions such as machining load and tool condition directly result in the change of cutting force. In particular, the real-time monitoring of the cutting force is instrumental in figuring out disturbances such as excessive machining load, unstable process conditions, and tool wear during the cutting process. In the thesis, the measurement results are utilized to describe the various phenomena related to the cutting process. For example, the mean value of the cutting forces is evaluated since the cutting force of elliptical vibration cutting is superior or lower than that of the ordinary cutting. As a different example, the frictional chatter or forced vibration causes the change of cutting forces. In addition, the period and amplitude of the dynamic forces due to the chip segmentation of the titanium alloy are observed.

A dynamometer is one of the standard means to measure the cutting forces directly. It consists of piezoelectric sensors clamped between two plates [11]. However, it is not desirable in general manufacturing environments since dynamometers are expensive and too sensitive to overload. There is often no space near the machining point for their installation. To overcome the limitations and to introduce the practical application on manufacturing machines, several studies have been conducted on indirect cutting force measurement [12]. For example, the current of the spindle drive system [13] or feed drive system [14], which are components of the machine tools, are used to estimate the cutting forces. Indirect cutting force measurements are applied to monitor the crucial parameters in the cutting process, such as the detection of tool breakage [15] and chatter [16]. In this work, a new method is proposed to detect the changes in machining load or tool wear in the elliptical vibration cutting process. The correlations between the change of internal data in the ultrasonic elliptical vibration cutting control system and the change of several parameters of the cutting process are demonstrated experimentally.

Vibration

During the cutting process, forced and self-excited vibrations often occur because of the cutting forces and the machine's structural dynamics. These vibrations cause problems such as the deterioration of the surface quality and short tool life. The self-excited vibration problem in the cutting process is called chatter, and it should be suppressed to obtain high performances of the cutting process [17]. Because the generation mechanisms of the chatter are different depending on the cutting process, suppression methods for those vibrations are also various [18]. Therefore, the data acquisition of the cutting process and machine tools and a comprehensive understanding of the cutting process and machine's structural dynamics are necessary. It is because the chatter is an excited vibration of the structural modes of the machine tool-workpiece system by cutting forces [19]. In this work, a kind of self-excited vibration, i.e., frictional chatter, in elliptical vibration cutting during the high-efficiency machining, that causes deterioration of the surface quality is investigated.

The transfer functions of the vibration device are measured through the hammering test. Then, the tool's chatter vibration is measured by a displacement sensor, and the frequency of chatter is calculated through the FFT analysis of the signals. As a result, it is possible to know which mode of the vibration tool is excited during the cutting process. Its growing mechanism is also explained by the cutting mechanics. The phase between the velocity of the second bending mode and the ploughing force is the reason for the chatter generation. Therefore, it is possible to suppress chatter by changing the lead angle. In addition, during the titanium alloy machining, the forced vibration due to the chip segmentation resulted in the increased cutting forces and deterioration of surface quality. From the photographs of the chips, the periodicities of the chip segmentation are calculated at different cutting conditions and compared with the results from the vibration wave of cutting forces and cut surface. As a result, the forced vibration is related to the cutting conditions such as average uncut chip thickness and cutting width. It is possible to suppress the forced vibration without sacrificing the machining efficiency by increasing the cutting width and decreasing the average uncut chip thickness.

Cut Surface and Tool Wear

Observations of the cut surfaces and tool wear are important for the evaluation of the cutting process. It is because the surface quality is one of the most specified customer requirements, and tool wear is an additional parameter which affects the surface quality of finished parts as well as the manufacturing cost. Therefore, many studies have been carried out to improve the cut surface quality and decrease the tool wear [20]. For example, to achieve ultra-precision surface quality, the diamond cutting is utilized [21], and to decrease the excessive tool wear, the elliptical vibration cutting is applied [22]. In this thesis, cutting experiments with long cutting distance is carried out to clarify the feasibility of the newly developed high-power elliptical vibration cutting device. After the experiments, the surface roughness of Ra and PV values are measured as indices of the surface quality of the finished surface. Tool wear is observed through the optical microscope. Furthermore, the cut surfaces are crucial evaluation elements to determine whether the undesirable vibration is imposed or suppressed in both frictional chatter and forced vibration due to chip segmentation in the thesis.

2.2 Real-World Data Analysis and Implementation in Cutting Process

To overcome the limitations of the conventional elliptical vibration cutting device, the high-power ultrasonic elliptical vibration cutting device is developed. Its feasibility is confirmed from several cutting experiments results. Here, cutting forces, surface roughness, and tool wear data are utilized. Furthermore, frictional chatter became the problem during the high-efficiency mirror surface finishing with elliptical vibration cutting. Hammering tests and cutting tests under various cutting conditions are conducted. The transfer function of the developed high-power ultrasonic elliptical vibration is calculated and dynamic properties of the device is verified. The vibration and cutting force data are analyzed in the frequency domain, and the frequency components of the chatter vibration and elliptical vibration are compared with the vibration marks on cut surfaces. Moreover, the time-domain simulation is also carried out by using the experimental results and clarified that the reason for the vibration marks on cut surfaces is the second resonant mode of bending vibration. Then, based on the model of the elliptical vibration cutting process, the ploughing process excites the second bending mode. As a result, its generation mechanism is confirmed, and the proper suppression method can be suggested. The verification test is also carried out, and with the collected data from the test, it can be concluded that the proposed suppression method is effective in suppressing the frictional chatter in high-efficiency elliptical vibration cutting. Figure 2 shows an illustration of the RWDC regarding chatter suppression for high-efficiency mirror surface finishing mentioned above.

The ultrasonic elliptical vibration cutting of titanium alloy is investigated. During the machining of titanium alloy, the forced vibration due to the segmented chip formation causes the deterioration of the cut surface and dynamic cutting forces. To suppress the forced vibration, several cutting experiments are conducted at various conditions. Time distribution data of cutting forces and vibration and image data of cut surface and chip are utilized. From the forced vibration analysis results, the relations between the segmented chip formation and the uncut chip geometry are identified. Based on the identified segmented chip generation mechanisms, it is possible to propose a practical and straightforward strategy to suppress forced vibration. As a result, it is concluded that the ultrasonic elliptical vibration cutting is an effective technology for reducing the cutting force, and precision machining is possible by suppressing the forced vibration. Figure 3 shows an illustration of the RWDC regarding forced vibration suppression caused by segmented chip formation for precision machining of titanium alloy mentioned above.

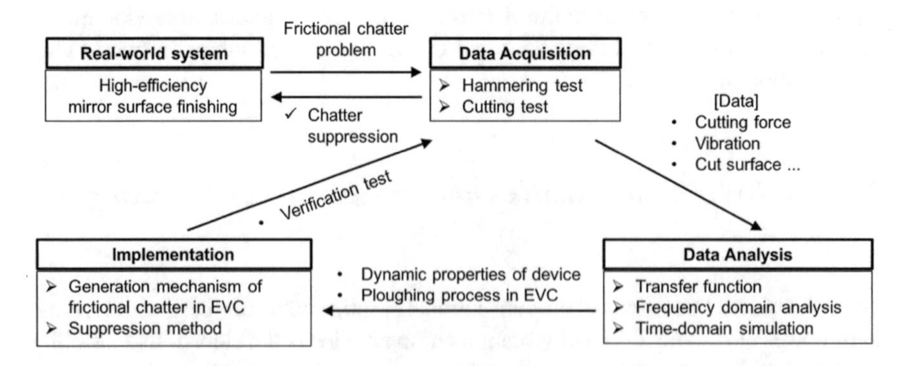

Fig. 2 Chatter suppression for high-efficiency mirror surface finishing

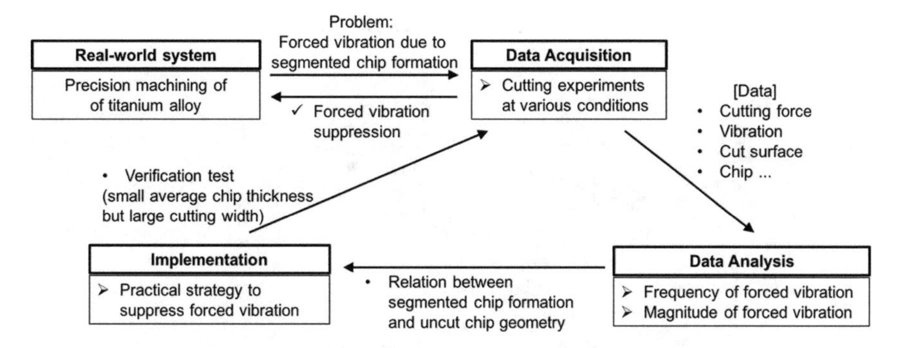

Fig. 3 Forced vibration suppression for precision machining of titanium alloy

A new method for monitoring of the elliptical vibration cutting process is proposed and experimentally verified. The internal data, such as currents, voltages, and excitation frequency, are measured from the elliptical vibration cutting tool controller. While the vibration status is changed according to the cutting process, those changes are reflected in the change of measured internal data. Based on the proposed vibration model of elliptical vibration cutting, the relations between the change of internal data and cutting process parameters, e.g., machining load and tool wear, are identified. It is expected that the process monitoring function investigated in this work can be utilized for process monitoring and automation without external sensors in the cutting process. Figure 4 shows an illustration of the RWDC regarding cutting process monitoring during the elliptical vibration cutting mentioned above.

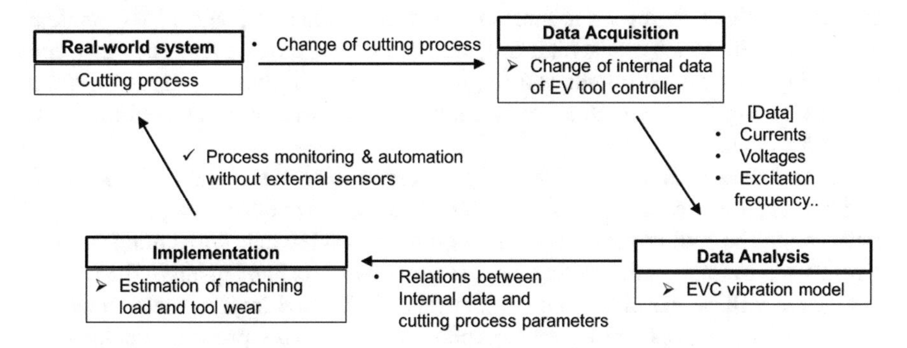

Fig. 4 Cutting process monitoring through internal data from EV tool controller

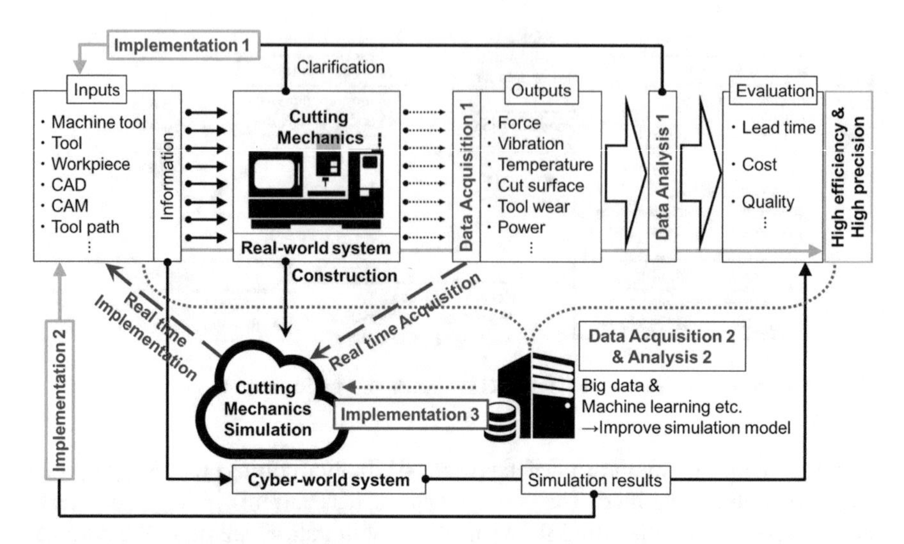

Fig. 5 Illustration of RWDC for advanced manufacturing process

2.3 *Real-World Data Circulation for Advanced Manufacturing Process*

Figure 5 shows the flow chart of the advanced manufacturing process with the Cyber-Physical System (CPS) for the perspective of promising Real-World Data Circulation. The author is confident that this RWDC will stimulate the successful realization of the CPS and the Industry 4.0-based manufacturing systems. The cutting process is considered as a multi-input and multi-output system. There is plenty of information as input variables such as machine tool, tool, workpiece, CAD, CAM, tool path, and machining conditions. Simultaneously, there is plenty of data as output variables such as cutting force, vibration, temperature, cut surfaces, wear of used tool, and power for the machine tool.

"Data Acquisition 1" describes the measurement or observation of the output parameters from the cutting process. "Data Analysis 1" indicates the procedure for interpreting the cutting mechanics based on the measured data and cutting process physics and evaluating the cutting process in terms of cost and quality. The cutting mechanics will be clarified from the analysis results, and improvement strategies of the cutting process can be set up and carried out. This process is defined as "Implementation 1" in Fig. 5. This data circulation is usually conducted by trial and error, which means many pieces of workpieces and tools are consumed until the feasibility is confirmed to be adopted in the actual manufacturing site.

Meanwhile, with the clarified cutting mechanics in the real-world system, it is possible to construct the "Cutting Mechanics" in the cyber-world system. The evaluation algorithms [23] of cutter-part-geometry intersection are developed and used to predict cutting forces, torque, power, and the possibility of having chatter by combining with cutting process physics and structural characteristics of the machine tool. Usually, there are recurrences of the trial and error cycle between the CAM data and real-world machining until it is competent. However, it is still not optimal. The simulation of cutting mechanics in digital environments can reduce or even eliminate physical trials in the real-world system. It can be used to search for optimal cutting conditions such as feed rate and spindle speed, rearrange the toolpath according to the optimal path strategy [24], and predict and avoid the chatter [25]. In this way, it is possible to reduce the cycle time, increase the cut surface qualities, and decrease the manufacturing costs [26].

"Real-time acquisition" and "Real-time implementation" indicate the integration of simulation of cutting mechanics with real-time measurements at the real-world systems [27]. For example, a virtually assisted online milling process control and monitoring system is proposed. It is utilized to avoid false tool failure detection and transient overloads of the tools during adaptive control [28]. To realize the CPS in the manufacturing site, this kind of integration is essential, and the accuracy and effectiveness of the process monitoring and cutting mechanics simulation should be advanced.

"Data Acquisition 2 and Analysis 2" indicates the procedures where the entire manufacturing process information and its data circulation are stored at every second and analyzed with advanced information technology. Most researches mentioned above are based on cutting process physics. Mathematically modeling the workpiece-tool engagement geometry and calculating the structural properties of the kinematic motion of the machine and workpiece to simulate the cutting mechanics in the cyber-world system are required for most of the research. However, there are also other researches based on statistical approaches. For example, neural network models or regression models are utilized to optimize the cutting conditions [29] or to predict the cut surface quality and tool wear [30]. Experimental results of the surface roughness and tool wear at different cutting parameters such as cutting speed, feed rate, and depth of cut are used as training data to construct the neural network models or regression models. There are many machining tools in the manufacturing site, and various cutting processes are conducted inside the machining tool continuously. Therefore, there is a tremendous amount of data, such as from the input and output variables of the cutting process to the evaluation elements of the cutting process, i.e., lead time, cost, and quality. It will be an important training data to construct a new or advanced cutting mechanics simulation ("Implementation 3"). Initially established cutting mechanics simulation is constructed based on the data acquired from the restricted situations of cutting experiments. Therefore, when an unforeseen situation occurs, the simulation cannot adaptively respond to new situations. However, by using the collected data and results of analysis from "Data Acquisition 2 and Analysis 2", more precise cutting mechanics simulation can be constructed. It will then improve the effectiveness and accuracy of the cutting mechanics in the real-world system, too. Moreover, these advanced data acquisition, analysis, and implementation will be applied to real-time or short-term improvements of the cutting process and long-term improvements such as the maintenance of the machine tool and index for new machine tool development.

3 Conclusion

Based on the researches in the work, the low-cost and high-efficiency mirror surface finishing of hardened steel, high-efficiency and high-precision machining of titanium alloy with low cutting forces, and a new function of the elliptical vibration cutting process monitoring can be achieved. Hence, this study is significantly advantageous to expand the applicable industrial fields because of its verified high performance and high precision. Furthermore, the applied Real-World Data Circulation, i.e., data acquisition, analysis, and implementation, in this work is categorized and described as well.

References

1. Jung H, Shamoto E, Ueyama T, Hamada S, Xu L (2016) Mirror surface finishing of hardened die steel by high-Power ultrasonic elliptical vibration cutting. J Mach Eng16(1):5–14
2. Jung H, Hayasaka T, Shamoto E (2016) Mechanism and suppression of frictional chatter in high-efficiency elliptical vibration cutting. Ann CIRP65(1):369–372
3. Jung H, Hayasaka T, Shamoto E (2018) Study on process monitoring of elliptical vibration cutting by utilizing internal data in ultrasonic elliptical vibration device. Int J Precis Eng Manuf -Green Technology 5(5):571–581
4. Jung H, Hayasaka T, Shamoto E, Xu L (2020) Suppression of forced vibration due to chip segmentation in ultrasonic elliptical vibration cutting of titanium alloy Ti–6Al–4V. Precis Eng 64:98–107
5. Wang S, Wan J, Zhang C (2015) Implementing smart factory of Industrie 4.0: an outlook. Int J Distrib Sensor Netw 12(1):1–10
6. Shrouf F, Ordieres J, Miragilotta G (2014) Smart factories in Industry 4.0: a review of the concept and of energy management approached in production based on the Internet of Things paradigm. Proceedings 2014 IEEE international conference on industrial engineering and engineering management, Bandar Sunway, 697–701
7. Bagheri R, Gill H (2011) Cyber-physical systems. Impact Control Technol 1(6):1–6
8. Lee J, Bagheri B, Kao HA (2015) A Cyber-Physical Systems architecture for Industry 4.0-based manufacturing systems. Manuf Lett 3:18–23
9. Kang HS, Lee JY, Choi SS, Park JH, Son JY, Kim BH, Noh SD (2016) Smart manufacturing: past research, present findings, and future directions. Int J Precis Eng Manuf -Green Technology 3(1):111–128
10. Ahmadi A, Cherifi C, Cheutet V, Ouzrout Y (2017) A review of CPS 5 components architecture for manufacturing based on standards. Proceedings 11th international conference on software, knowledge, information management and applications (SKIMA), Malabe, 1–6
11. Gautschi GH (1971) Cutting forces in machining and their routine measurement with multi-component piezo-electric force transducers. In: Koenigsberger F, Tobias SA (eds) Proceedings of 12th international machine tool design and research conference, Palgrave, London, 113–120
12. Li X (2005) Development of current sensor for cutting force measurement in turning. IEEE Trans Instrum Measur 54(1):289–296
13. Chang YC, Lee KT, Chuang HY (1995) Cutting force measurement of spindle motor. J Contr Syst Technol 3(2):145–152
14. Altintas Y, Dong CL (1990) Design and analysis of a modular CNC system for machining control and monitoring. In: Ferreira PM, Kapoor SG, Wang ACY (eds) Modeling of machine tools: accuracy, dynamics, and control—american society of mechanical engineering 45:199–208

15. Lee JM, Choi DK, Chu CN (1995) Real-time tool breakage monitoring for NC milling process. Ann. CIRP 44(1):59–62
16. Smith S, Tlusty J (1993) Stabilizing chatter by automation spindle speed regulation. Ann CIRP 41(1):433–436
17. Munoa J, Beudaert X, Dombovari Z, Altintas Y, Budak E, Brecher C, Stepan G (2016) Chatter suppression techniques in metal cutting. Ann CIRP 65(2):785–808
18. Altintas Y, Budak E (1995) Analytical prediction of stability lobes in milling. Ann CIRP 44(1):357–362
19. Altintas Y (2000) Manufacturing automation: metal cutting mechanics, machine tool vibrations, and CNC design. Cambridge University Press
20. Shamoto E, Suzuki N (2014) Ultrasonic vibration diamond cutting and ultrasonic elliptical vibration cutting. Comprehensive Mater Process 11:405–454
21. Moriwaki T (1989) Machinability of copper in ultra-precision micro diamond cutting. Ann CIRP 38(1):115–118
22. Shamoto E, Moriwaki T (1994) Study on elliptical vibration cutting. Ann CIRP 43(1):35–38
23. Spence A, Altintas Y (1991) CAD assisted adaptive control for milling. Trans ASME J Dynamic Syst Measur Contr 113:444–450
24. Lazglu I, Manav AC, Murtezaoglu Y (2009) Tool path optimization for free form surface machining. Ann CIRP 58(1):101–104
25. Eksioglu C, Kilic ZM, Altintas Y (2012) Discrete-time prediction of chatter stability, cutting forces, and surface location errors in flexible milling systems. Trans. ASME J Manuf Sci Eng 134(5):AN071006
26. Altintas Y, Kersting P, Biermann D, Budak E, Denkena B, Lazoglu I (2014) Virtual process systems for part machining operations. Ann CIRP 63(2):585–605
27. Liang SY, Hecker RL, Landers RG (2002) Machining process monitoring and control: the state-of-the-art. Proceedings ASME 2002 international mechanical engineering congress and exposition, 599–60
28. Altintas Y, Aslan D (2017) Integration of virtual and on-line machining process control and monitoring. Ann CIRP 66(1):349–352
29. Zuperl U, Cus F (2003) Optimization of cutting conditions during cutting by using neural networks. Robot Computer Integrated Manuf 19:189–199
30. Asilturk I, Cunkas M (2011) Modeling and prediction of surface roughness in turning operations using artificial neural network and multiple regression method. Expert Syst Appl 38:5826–5832

A Study on Efficient Light Field Coding

Kohei Isechi

Abstract This section introduces a study on light field coding as one example of real-world data circulation. A light field is a set of dense multi-view images of which intervals between viewpoints are quite small. The light field has rich 3D visual information, so that the light field enables many kinds of 3D image processing. Nowadays, the light field has become one of the fundamental data formats for 3D image processing. The light field used for the practical applications consists of tens-to-hundreds images, so that compression of a light field is required. This section introduces a light field coding scheme of approximating a light field with the sum of weighted binary patterns. At the end of this section, this study is reviewed in the light of data circulation.

1 Introduction

This section introduces a study on light field coding using weighted binary patterns as one example of real-world data circulation. A light field is a four-dimensional (4D) signal describing light rays which travel in three-dimensional (3D) space [6, 10, 19]. The appearance of the world can be visually observed by receiving the light rays which reach our eyes; therefore, visual information can be described by formulating the light rays. The 4D light field signal can be interpreted as a set of dense multi-view images of which intervals between viewpoints are quite small. Figure 1 shows examples of still light field images with different number of viewpoints. The dense multi-view images can be captured by using special camera systems such as multi-camera array systems [7, 32] and a light field camera [1, 9, 22, 23]. The captured multi-view images are a kind of real-world data. Since the light field signal includes rich visual information in the 3D space, the light field data can be utilized for various kinds of applications such as depth estimation [12, 26, 27, 30], free-viewpoint image

K. Isechi (✉)
Graduate School of Engineering, Nagoya University, Furo-cho, chikusa-ku,
Nagoya, Aichi 464-8601, Japan
e-mail: isechi@fujii.nuee.nagoya-u.ac.jp

K. Takeda et al. (eds.), *Frontiers of Digital Transformation*,
https://doi.org/10.1007/978-981-15-1358-9_7

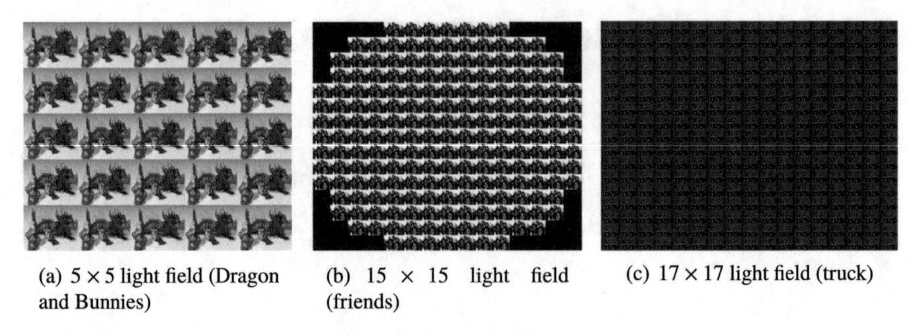

(a) 5×5 light field (Dragon and Bunnies) (b) 15×15 light field (friends) (c) 17×17 light field (truck)

Fig. 1 Light fields with different number of viewpoints

synthesis [4, 19, 25], and stereoscopic displaying [17, 18, 24, 31]. Nowadays, the light field has become one of the fundamental data formats for 3D image processing.

One of the important issues for handling the light field is the huge data size. Light fields used in the practical applications consist of tens-to-hundred multi-view images. Light fields with a larger number of viewpoints are desirable to provide high-quality 3D experiences in the practical applications of a light field described above. Increasing the number of viewpoints leads to an increase in the data amount of a light field. In the case of light field videos, multi-view images are accumulated in time sequence; consequently, its data amount becomes huge with time.

Due to the huge data size of a light field, efficient coding is one of the important research topics for handling a light field. The huge data size of a light field causes difficulty in light field applications. For example, assuming a light field streaming system, it would occupy much larger network bandwidth than a general single-viewpoint video streaming system; thus, efficient light field coding scheme is required. The light field has large redundancy which is different from typical single-viewpoint video because the light field images are similar to each other due to quite small intervals between viewpoints as shown in Fig. 1. By extracting the redundancy, the amount of light field data can be dramatically reduced. There are many researches which focus on efficiently removing the redundancy of a light field.

Most research uses modern video coding standards such as H.265/HEVC [13] with small modifications [11, 15, 20, 21, 28, 29]. The modern video coding standards have been developed for compressing typical 2D videos where various kinds of techniques such as intra/inter-frame prediction, discrete cosine transform (DCT), and arithmetic coding are employed to eliminate the redundancy of the 2D video sequences. A light field can be regarded as a 2D video sequence by aligning viewpoint images in some order, and the modern video coding standards can be applied to the light field. In particular, motion compensation techniques, which predict motion of captured objects between sequential frames in the target video, are useful for removing redundancy among the images at different viewpoints in a dense light field. The modern coding standards with small modifications have achieved good coding performance for a light field.

As a novel approach for light field coding, Fujii Laboratory in Nagoya University has proposed a light field coding scheme with weighted binary patterns in [16]. In contrast to the coding methods using the modern video coding standards with small modifications, this scheme has been developed based on a mind that the standard video coding techniques are not necessarily the most suitable for a dense light field. The key idea of this scheme is that a light field is approximated with only several weighted binary patterns. The binary patterns and weight values are computationally obtained so as to optimally approximate the target light field. This scheme is named "baseline" as a counterpart of the progressive scheme mentioned later.

The baseline coding scheme is completely different from those of modern video coding standards, and its decoding process is dramatically simpler than that of the standard codecs. The simplicity of decoding process allows us to implement faster and less power-hungry decoders than those of the standard codecs. However, the encoding process, i.e., solving the optimization problem, takes much longer than the video coding standards as the number of binary patterns increases. To accelerate the encoding process, a progressive framework that progressively approximates a light field with a small number of weighted binary patterns at each group has been proposed. The progressive framework accelerates the encoding process of the baseline method while keeping feasible accuracy, but causes a problem of the trade-off between computational complexity and coding performance.

To address the trade-off problem, disparity compensation framework has been introduced into the progressive method [14]. Disparity compensation shifts the pixels in the images according to a specified disparity value and the viewpoint positions. This section briefly describes the framework of baseline, progressive, and progressive with disparity compensation methods, and demonstrates their performances.

The study described in this section has been undertaken based on the concept of real-world data circulation. In this case, light field data, namely a set of multi-view images, is real-world data describing visual information in the 3D world. The detail of data circulation structure is described at the end of this section.

1.1 Baseline Method

First, a framework of "baseline" light field coding is introduced. A light field to be compressed is given as $L(s, t, x, y)$. The set of symbols (s, t), $(s = 1, \dots, S, t = 1, \dots, T)$ and (x, y), $(x = 1, \dots, X, y = 1, \dots, Y)$ indicate viewpoint coordinates in a 2D grid and pixel coordinates in a viewpoint image, respectively. The baseline method approximates a light field $L(s, t, x, y)$ by using the sum of N weighted binary patterns as follows:

$$L(s, t, x, y) \simeq \sum_{n=1}^{N} B_n(x, y) r_n(s, t), \quad (1)$$

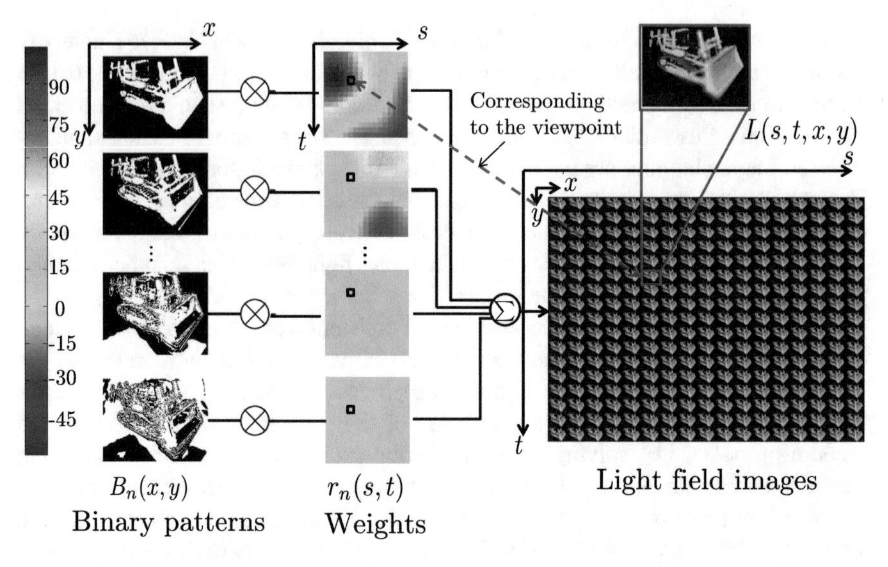

Fig. 2 Light field approximation using binary patterns $B_n(x, y)$ and corresponding weights $r_n(s, t)$

where $B_n(x, y) \in \{0, 1\}$, $(n = 1, 2, \ldots, N)$, $r_n(s, t) \in \mathbb{R}$, and b_L indicate binary patterns, corresponding weights, and a bit depth used for light field pixels, respectively. Figure 2 illustrates the framework of light field approximation using weighted binary patterns. All the images at different viewpoints are represented with the same binary patterns, but different weights are used depending on the viewpoints; therefore, the binary patterns and the weights have the common and viewpoint-dependent components of the multi-view images, respectively.

To obtain binary patterns and weights that can accurately approximate a given light field, the following optimization problem is solved.

$$\mathop{\arg\min}_{\substack{B_n(x,y)\\ r_n(s,t)}} \sum_{s,t,x,y} \|L(s, t, x, y) - \sum_{n=1}^{N} B_n(x, y) r_n(s, t)\|^2. \tag{2}$$

The optimal solution to Eq. (2) cannot be easily obtained because the equation includes two sets of unknowns, $B_n(x, y)$ and $r_n(s, t)$. An alternating optimization can be utilized for obtaining the solution, where the binary patterns $B_n(x, y)$ are initialized at first, and the following two steps are repeated until convergence: (i) the binary patterns are fixed, and the weights are optimized, (ii) the weights are fixed, and the binary patterns are optimized. The optimization of binary patterns (i) can be regarded as a standard least squares minimization problem that can be easily solved by well-known methods. The solution to the optimization of weights (ii) can be obtained individually for each pixel (x, y) because this problem is pixel-independent. The optimization (ii) is regarded as a binary combinational optimization known as an NP-hard problem. Its solution can be obtained by using a simple brute-force search although the computational complexity of brute-force search is heavy and increases exponentially as the number of binary patterns increase. After convergence, a light field $L(s, t, x, y)$ can be reconstructed using Eq. (1) with the obtained solution to Eq. (2). As one implementation, all $B_n(x, y)$ can be initialized to the same image obtained by applying binary thresholding to the top-leftmost image of the target light field. Since all the binary patterns are identical at this point, the weights $r_n(s, t)$ cannot be determined uniquely in the optimization (i) of the first iteration.

Similar to other coding schemes, the proposed scheme has a trade-off between rate (number of bits) and distortion (accuracy of decoded data). This trade-off can be controlled by simply changing the number of binary patterns N. When a target light field is represented by using N binary patterns and weights, the compression ratio of the scheme is calculated as follows:

$$\text{compression ratio} = \frac{N(XY + STb_r)}{STXYb_L}, \tag{3}$$

where b_r indicates bit depth used for weights $r_n(s, t)$. For instance, a compression ratio for light field images with $X = 160$, $Y = 120$, $S = 17$, $T = 17$, and $b_L = 8$ equals 0.54% in the case of $N = 10$ and $b_r = 16$. This compression ratio is calculated based on the binary patterns and weights which are not compressed. To further reduce the data amount, a lossless compression algorithm, e.g., gzip, can be applied to the binary patterns and weights.

An important advantage of the proposed scheme is extreme simplicity of decoding process. As shown in Fig. 2, the light field can be reconstructed by using only product-sum operations. On the other hand, the decoding process of modern video coding standards includes complicated operations such as intra/inter-frame prediction, inverse DCT/DST transformation, and dequantization. Thus, the decoding process of the proposed scheme is dramatically simpler than that of modern video coding standards. This simplicity allows us to implement a faster and less power-hungry decoder using field programmable gate array (FPGA) and an application specific integrated circuit (ASIC).

1.2 Progressive Method

Next, a framework of "progressive" light field coding, which accelerates the encoding process of baseline method, is expounded. To accelerate the encoding process, the coding scheme is made progressive on the basis of the divide-and-conquer strategy as shown in Fig. 3. Assuming that an original light field $L(s, t, x, y)$ is approximated by using totally N binary patterns and weights, the N binary patterns are divided into M groups (layers). Each layer has $\mathcal{N}$ binary patterns so that $\mathcal{N}M = N$ is satisfied. At the first layer, the target light field $L_1(s, t, x, y)$ is set as an original light field $L(s, t, x, y)$ and is approximated by using $\mathcal{N}$ binary patterns and weights based on Eq. (2). The light field obtained by the approximation is denoted as $L'_1(s, t, x, y)$.

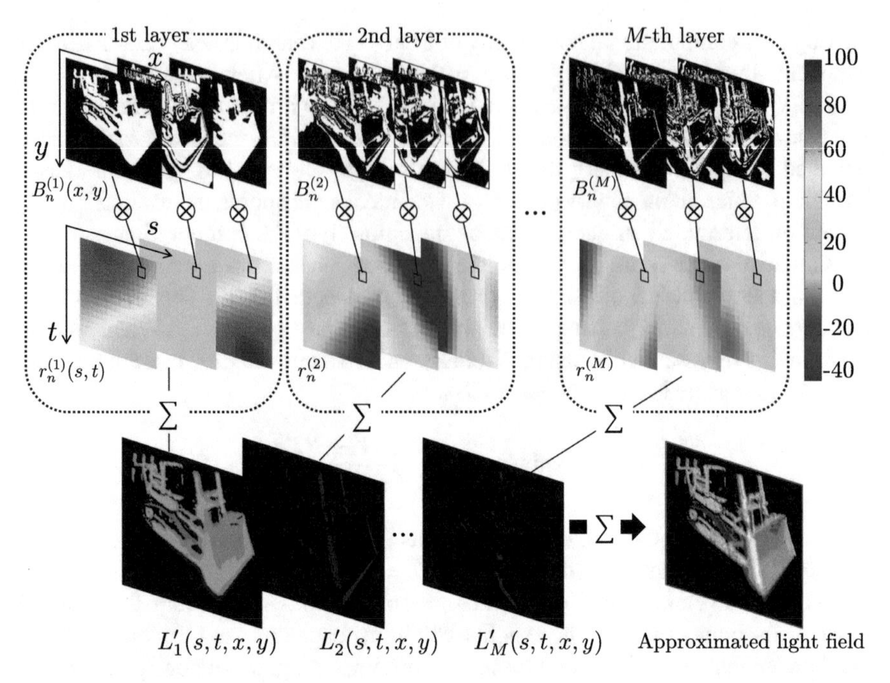

Fig. 3 Progressive coding framework

At the next layer, the target light field $L_2(s, t, x, y)$ is defined as the difference between $L(s, t, x, y)$ and $L'_1(s, t, x, y)$. The target light field $L_2(s, t, x, y)$, i.e., the residual light field is also approximated with $\mathcal{N}$ binary patterns and weights in the same manner as the first layer. At the following layers, the process is progressively repeated where $\mathcal{N}$ binary patterns and weights are calculated to approximate the target light field $L_m(s, t, x, y)$ as follows:

$$\mathop{\arg\min}_{\substack{B_n^{(m)}(x,y)\\ r_n^{(m)}(s,t)}} \sum_{s,t,x,y} \| L_m(s, t, x, y) - L'_m(s, t, x, y) \|^2, \tag{4}$$

$$L_m(s, t, x, y) = L(s, t, x, y) - \sum_{i=1}^{m-1} L'_i(s, t, x, y), \tag{5}$$

$$L'_i(s, t, x, y) = \sum_{n=1}^{\mathcal{N}} B_n^{(i)}(x, y) r_n^{(i)}(s, t), \tag{6}$$

where $B_n^{(i)}(x, y)$, $r_n^{(i)}(s, t)$, and $L'_i(s, t, x, y)$ indicate the n-th binary pattern, the corresponding weight, and the approximated light field at the i-th layer, respectively. Finally, an original light field $L(s, t, x, y)$ is approximated as follows:

$$L(s,t,x,y) \simeq \sum_{m=1}^{M} L'_m(s,t,x,y). \tag{7}$$

This progressive extension achieves remarkable reduction of the computational complexity by optimizing a small number of binary patterns and weights per layer on the basis of the divide-and-conquer strategy. Although the strategy generally cannot bring about the global-optimum solution, a feasible solution can be obtained with less computational complexity since the search space of the problem is reduced. In the progressive coding, dividing the binary patterns and weights into M groups is equal to the reduction of search space. The main bottleneck of the baseline scheme is optimizing binary patterns; thus, reducing the number of binary patterns in the optimization can significantly accelerate the encoding process. The computational complexity for binary combinational optimization in the baseline scheme with N binary patterns is $\mathcal{O}(2^N)$, but the computational complexity for that in the progressive scheme is reduced to $\mathcal{O}(2^{\mathcal{N}})$ by dividing N binary patterns into M layers. Total computational complexity for binary combinational optimization in the progressive scheme equals $\mathcal{O}(M2^{\mathcal{N}})$ since the progressive scheme repeats the optimization for each layer. For instance, the computational cost of the progressive coding is reduced to 1/128 compared with that of the baseline one in the case with $N = 12$, $M = 4$, and $\mathcal{N} = 3$. As described above, the progressive extension finds a feasible solution with less computational complexity, but the approximation accuracy cannot help being degraded since the divide-and-conquer strategy cannot find the global-optimum solution. Consequently, there is a trade-off between computational complexity and rate-distortion performance.

1.3 Disparity Compensation Framework for Binary Patterns

To improve the rate-distortion performance while avoiding infeasible computational complexity, a method of adapting a disparity compensation framework to the progressive light field coding scheme is introduced. In the conventional scheme, all the images at different viewpoints are approximated using the same binary patterns with different weights. This means that the binary patterns represent the common components among multi-view images and that the weights represent viewpoint-dependent components. Meanwhile, if disparity compensation is applied to the binary patterns, they can represent not only the common components but also the viewpoint-dependent components. Thus, applying appropriate disparity compensation might improve the representation capability of the binary patterns. The amount of disparities included in a light field depend on a captured scene; therefore, the appropriate disparity values should be adaptively searched for according to the scene. The contents of this section have been published in a journal paper [14].

With disparity compensation applied, the approximation of a light field can be formulated as follows:

$$L(s,t,x,y) \simeq \sum_{n=1}^{N} B_n(x - sd_n, y - td_n) r_n(s,t), \tag{8}$$

where d_n is a disparity value with which the n-th binary pattern is compensated. In the case where $x - sd_n$ and $y - td_n$ become non-integer values, pixel values at integer position are interpolated by using the neighboring pixel values existing at non-integer position. If $x - sd_n$ and $y - td_n$ indicate the out of an image, as one approach, it can be assumed that the pixel value at the corner position continuously exists at the out of an image although there are several approaches for dealing with this case. All pixels of $B_n(x, y)$ are shifted according to the viewpoint position (s, t) so that each viewpoint image is approximated by using slightly shifted binary patterns depending on the viewpoint position. According to Eq. (8), the optimization problem to find the binary patterns, weights, and disparity value is defined as follows:

$$\mathop{\arg\min}_{\substack{B_n(x,y) \\ r_n(s,t) \\ d_n}} \sum_{s,t,x,y} \|L(s,t,x,y) - L'(s,t,x,y)\|^2, \tag{9}$$

$$L'(s,t,x,y) = \sum_{n=1}^{N} B_n(x - sd_n, y - td_n) r_n(s,t). \tag{10}$$

Although it would be preferable if the global-optimum solution for Eq. (9) can be obtained, solving this optimization is quite difficult because it includes three sets of unknowns. Eq. (8) is reformulated with a restriction on disparity values d_n and then introducing a disparity compensation framework into the progressive light field coding scheme presented in Sect. 1.2.

The progressive light field coding with disparity compensation is formulated as follows:

$$L(s,t,x,y) \simeq \sum_{m=1}^{M} L'_m(s,t,x,y) \tag{11}$$

$$L'_m(s,t,x,y) = \sum_{n=1}^{N} B_n^{(m)}(x - sd_m, y - td_m) r_n^{(m)}(s,t), \tag{12}$$

where d_m is a disparity value used for the disparity compensation at the m-th layer. As a restriction, only one disparity value is used for each group of binary patterns. At each layer, binary patterns, weights, and a disparity value are obtained by solving the following optimization:

$$\mathop{\arg\min}_{\substack{B_n^{(m)}(x,y) \\ r_n^{(m)}(s,t) \\ d_m}} \sum_{s,t,x,y} \|L_m(s,t,x,y) - L'_m(s,t,x,y)\|^2, \tag{13}$$

$$L_m(s,t,x,y) = L(s,t,x,y) - \sum_{i=1}^{m-1} L'_i(s,t,x,y). \quad (14)$$

Equation (13) still includes three sets of unknowns like Eq. (9). The range of disparities included in a dense light field is basically narrow because of its very small viewpoint interval; thus, the problem is simplified by manually defining a set of candidate disparities $\mathcal{D}$. The specific algorithm is shown in Algorithm 1, where $B_n^{(m)*}(x,y)$, $r_n^{(m)*}(s,t)$, and d_m^* denote a solution for Eq. (13) at the m-th layer. The binary patterns and weights are calculated for each candidate disparity value $d \in \mathcal{D}$. From the set of the obtained binary patterns and weights, the best one that achieves the best approximation accuracy is employed. Consequently, the proposed method adaptively searches for the appropriate disparity value at each layer depending on a captured scene by simple brute-force search.

To practically solve Eq. (13), it is reformulated as follows:

$$\mathop{\arg\min}_{\substack{B_n^{(m)}(x,y)\\ r_n^{(m)}(s,t)\\ d_m}} \sum_{s,t,x,y} \| L_m(s,t,x'+sd_m, y'+td_m) - L'_m(s,t,x'+sd_m, y'+td_m) \|^2, \quad (15)$$

$$L'_m(s,t,x'+sd_m, y'+td_m) = \sum_{n=1}^{N} B_n^{(m)}(x',y') r_n^{(m)}(s,t), \quad (16)$$

where $x' = x - sd_m$, and $y' = y - td_m$. The right side of Eq. (16) takes the same form as the second term in Eq. (2). Disparity compensation is first applied to the target light field $L_m(s,t,x,y)$ to obtain $L_m(s,t,x'+sd, y'+td)$; after that, the solution for (15) is obtained in the same manner as solving Eq. (2). The desired light field $L'(s,t,x,y)$ can be obtained by applying inverse disparity compensation to the light field $L'(s,t,x'+sd_m, y'+td_m)$.

By manually defining a set of candidate disparity values, the proposed method finds the best disparity values while keeping feasible computational complexity. The difference between the proposed scheme and the conventional progressive method with respect to computational complexity is the brute-force search for disparity values. Assuming that the number of elements of $\mathcal{D}$, i.e., the number of candidate disparity values, is denoted as D, the computational complexity for optimizing binary patterns in the proposed scheme is $O(DM \cdot 2^N)$ because the proposed method obtains the binary patterns and weights with all candidate disparity values at each layer. Although the proposed method takes much more time for encoding than the conventional progressive coding, it still can find the solution with feasible computational complexity. As mentioned in Sect. 1.2, the main bottleneck is optimizing binary patterns; thus, if N is kept small, the computational complexity of the proposed method never increases explosively like that of the conventional baseline scheme as the total

Algorithm 1 Progressive light field coding using disparity-compensated and weighted binary patterns

Input: $L(s, t, x, y)$, $\mathcal{D}$
Output: $B_n^{(m)*}(x, y), r_n^{(m)*}(s, t), d_m^*$ $(n = 1, \cdots, N, m = 1, \cdots, M)$
Initialize $L_1(s, t, x, y) \Leftarrow L(s, t, x, y)$
for $i = 1$ to M **do**
  BEST_PSNR $\Leftarrow$ 0.0
  for each disparity $d \in \mathcal{D}$ **do**
    Obtain $B_n^{(i)}(x, y), r_n^{(i)}(s, t)$ using Eq. (13) with fixed $d_m = d$
    $p \Leftarrow$ PSNR of $L(s, t, x, y)$ from Eq. (11) with $M = i$
    if BEST_PSNR $< p$ **then**
      $B_n^{(i)*}(x, y) \Leftarrow B_n^{(i)}(x, y)$
      $r_n^{(i)*}(s, t) \Leftarrow r_n^{(i)}(s, t)$
      $d_i^* \Leftarrow d$
      BEST_PSNR $\Leftarrow p$
    end if
  end for
  Carry over the residual using Eq. (14) with $m = i + 1$
  $i \Leftarrow i + 1$
end for

number of binary patterns N increases. In the case with $N = 24$, $M = 8$, $\mathcal{N} = 3$, and $D = 10$, the proposed scheme takes 10 times longer than the conventional progressive scheme, but the computational cost of the proposed scheme $10 \times 8 \times 2^3 = 640$ is still feasible compared with that of the baseline scheme 2^{24}.

When a target light field is represented by using N binary patterns and the corresponding weights, which are divided in M groups, the compression ratio of the proposed scheme is calculated as follows:

$$\text{compression ratio} = \frac{N(XY + STb_r) + Mb_d}{STXYb_L}, \tag{17}$$

where b_d denotes a bit depth used to describe the used disparity values. The increase of total bits compared with Eq. (3) is less of an issue because one byte ($b_d = 8$) is enough to describe a disparity value when D is set to 10–20.

1.4 Experimental Results

To evaluate the performance of the coding schemes using weighted binary patterns, the baseline method, progressive method, and progressive with disparity compensation method were implemented using a software made available from the website [8]. Six light field datasets [3] shown in Fig. 4, each of which consists of 17×17 grayscale multi-view images, were used in the experiments. For optimizing Eqs. (2) and (6), the number of iterations was set to 20. All binary patterns $B_n(x, y)$ were initialized

Datasets	Truck	Bulldozer	Amethyst
# of views		17 × 17	
size	160 × 120	192 × 144	96 × 128
Top-left image			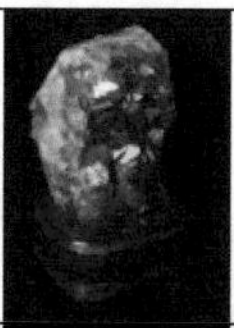
Difference between top-left and top-right images		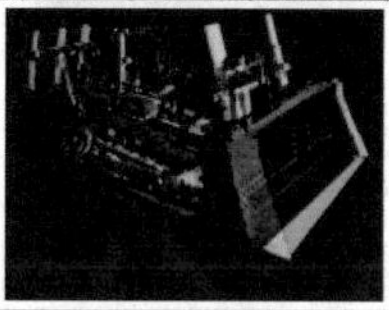	

Datasets	Bunny	Crystal	Knight
# of views		17 × 17	
size	128 × 128	128 × 128	128 × 128
Top-left image			
Difference between top-left and top-right images			

Fig. 4 Datasets

by the result of binary thresholding to the most top-left image of the input light field. To compress the binary patterns and weight values, gzip ver.1.6 was adopted as an optional post-processing. Two implementations were adapted for H.265/HEVC: FFmpeg ver. 4.1 with default parameters and the HEVC Test Model [5] with a random access configuration [2]. As the difference between FFmpeg and HEVC Test Model, FFmpeg makes use of the x265 library for HEVC encoding and its default option is chosen to provide moderate coding efficiency with a reasonable complexity, while the configuration of the HEVC Test Model is basically determined to bring out the potential performance in the standardization process. To ascertain the effect of inter-frame prediction, all intra mode with the HEVC Test Model were also tested. When these video codecs were applied to a light field dataset, images in the dataset were aligned in the row-major order and were regarded as a single-viewpoint video sequence.

First, the rate-distortion performance and encoding time of the baseline and the progressive methods were evaluated. The PSNR values were calculated from the mean squared errors over all the viewpoints and pixels. The bitrate values were calculated from the sizes of files that were compressed with gzip. Computational

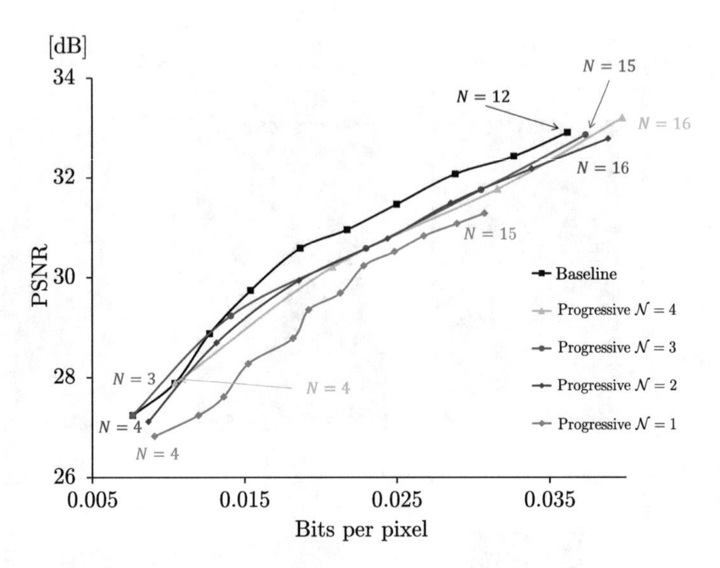

Fig. 5 Rate-distortion performances of baseline and progressive methods

Table 1 Encoding time ($N = 12$)

Method	Time (s)
Progressive $N = 1$	51
Progressive $N = 2$	54
Progressive $N = 3$	66
Progressive $N = 4$	87
Baseline	8533

times were measured on a Desktop PC running Windows 10 Pro equipped with Intel (R) Core (TM) i5-4590 3.3-GHz CPU and 8.0 GB main memory.

Figure 5 [16] shows that rate-distortion curves of the baseline and the progressive method. As shown in Fig. 5, using the progressive framework slightly decreases the coding efficiency compared to the baseline one. Table 1 shows the encoding time with and without the progressive framework, where the total number of binary images N was fixed to 12. Table 1 demonstrates that an approximately 100-fold increase in speed is achieved with the progressive framework. In terms of the balance between the rate-distortion performance and the encoding time, $N = 3$ is likely to be the best parameter in the progressive framework. Therefore, this configuration is employed from the following evaluation.

As shown in Fig. 6, the baseline and progressive methods were compared with H.265/HEVC in regard to the rate-distortion performances over several grayscale datasets taken from [3]. Performances of the baseline and progressive method appear to depend on the image differences between the viewpoints. They achieved a good rate-distortion performance for a dataset with small differences, such as Amethyst,

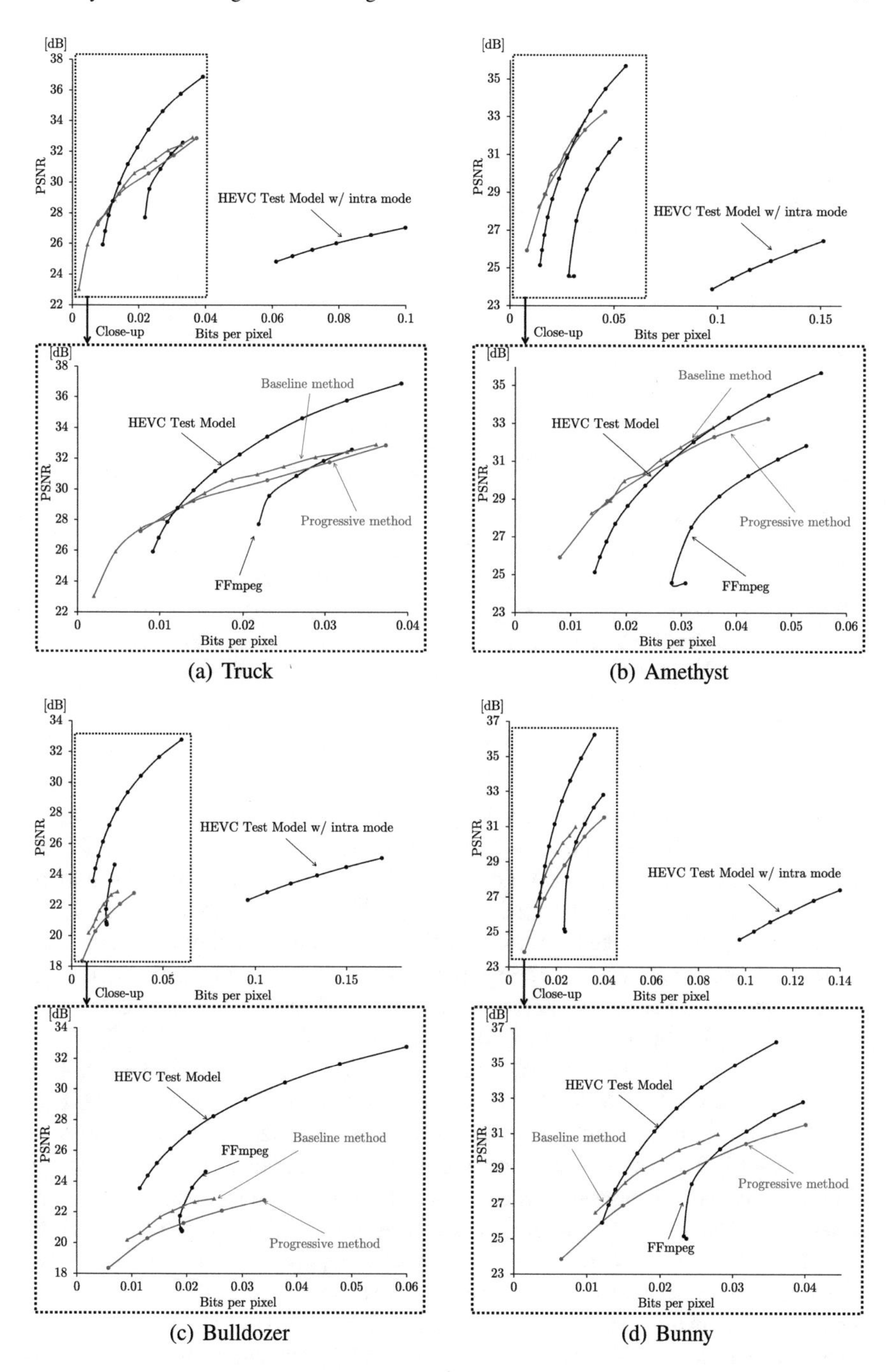

Fig. 6 Comparison of R-D curves

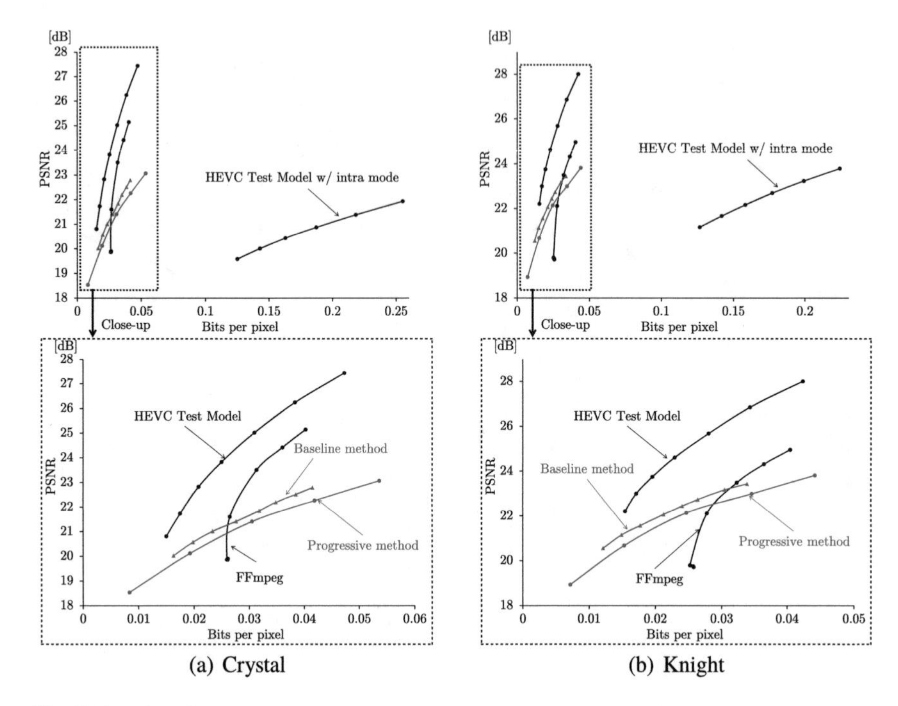

(a) Crystal (b) Knight

Fig. 6 (continued)

but could not achieve high PSNR values for a dataset including large differences, such as Bulldozer. They achieved a reasonable performance overall, comparable to that of FFmpeg but moderately inferior to that of the HEVC Test Model. The authors believe that these results are promising because HEVC is the state-of-the-art video coding standard that has been optimized with a significant amount of labor and time.

Next, the decoding times were compared between the baseline method and HEVC codecs using the truck dataset in grayscale. The Desktop PC running Windows 10 pro equipped with Intel (R) Core (TM) i5-4590 3.3-GHz CPU and 8.0 GB main memory was used, but the decoding time was measured on Ubuntu 16.04 installed on Virtual Box ver 5.1.16 due to the availability of software. The user time was used for executing decoding processes with "time" command. The measurement was repeated 100 times and obtained the average for each method, which is plotted in Table 2 [16]. For the baseline method, N was set to 10 and the outputs were written as pgm files. The time for unzip process was negligible. For FFmpeg and the HEVC Test Model, the outputs were, respectively, written in pgm files and YUV files. The parameters for these codecs was set so that they resulted in almost the same PSNR as that of the scheme using weighted binary patterns. It can be seen from the graph that the method with weighted binary patterns runs much faster than HEVC codecs. This can be attributed to the simplicity of the scheme; as mentioned earlier, its decoding process is carried out with simple sum-of-product operations, while HEVC requires complex inter/intra-frame prediction and transforms.

Table 2 Comparison of decoding time

Method	Time [s]
HEVC test model	0.158
FFmpeg	0.064
Baseline method	0.012

The effectiveness of the disparity compensation framework in the progressive light field coding scheme was investigated on the basis of rate-distortion performance. The set of candidate disparity values $\mathcal{D}$ in the progressive with disparity compensation method was given as $\mathcal{D} = \{0.0, \pm 0.2, \pm, 0.5, \pm 0.8, \pm 1.0, \pm 1.5, \pm 2.0\}$; namely, the number of candidate disparity values is $D = 13$. The proposed method, the conventional progressive coding, and the baseline coding were compared. The number of binary patterns N for the proposed method and the conventional progressive coding was varied from 3 to 24, but N for the baseline coding was limited from 3 to 12 because of the high computational complexity. The number of binary patterns in each layer was set as $\mathcal{N} = 3$. The bitrate was calculated from raw binary patterns, weights, and used disparity values without gzip compression.

Figure 7 shows rate-distortion curves for six datasets. The conventional baseline and progressive methods are called "Baseline" and "Progressive", respectively, and the proposed method is called "Progressive + disp. comp." in the results. Compared to the conventional progressive coding, the proposed method remarkably improves rate-distortion performances for truck and bulldozer while showing almost the same performances for the other datasets. The rate-distortion performances of the proposed method for truck and bulldozer even outperform those of the baseline method. Table 3 indicates selected disparity values in the proposed method with $N = 24$ for each layer of six datasets. The proposed method finds and uses non-zero disparity values in most of the layers for truck and bulldozer; thus, the disparity compensation outstandingly makes a difference for the two datasets. Meanwhile, for the other datasets, the selected disparities were zero for almost all the layers, which explains the reason why the progressive method with and without disparity compensation performed similarly in Figs. 7b–f; these datasets have only small disparities, and thus, the progressive method without disparity compensation was sufficiently effective for them. Figure 8 presents visual comparisons between the proposed method and the conventional progressive coding. It seems that the proposed method achieves better approximation accuracy by alleviating blurs on the object's parts having large disparities.

Table 4 shows the comparison of encoding time of three methods for truck. The encoding time was measured on a desktop PC running Windows 10 Pro equipped with an Intel Core (TM) i7-6700 3.4-GHz CPU and 16.0-GB main memory. The proposed method takes much more time for encoding than the conventional progressive method. However, the encoding time of the proposed method linearly increases as the number of binary patterns increases, while the encoding time of the baseline method explosively increases. The experimental results prove that the proposed

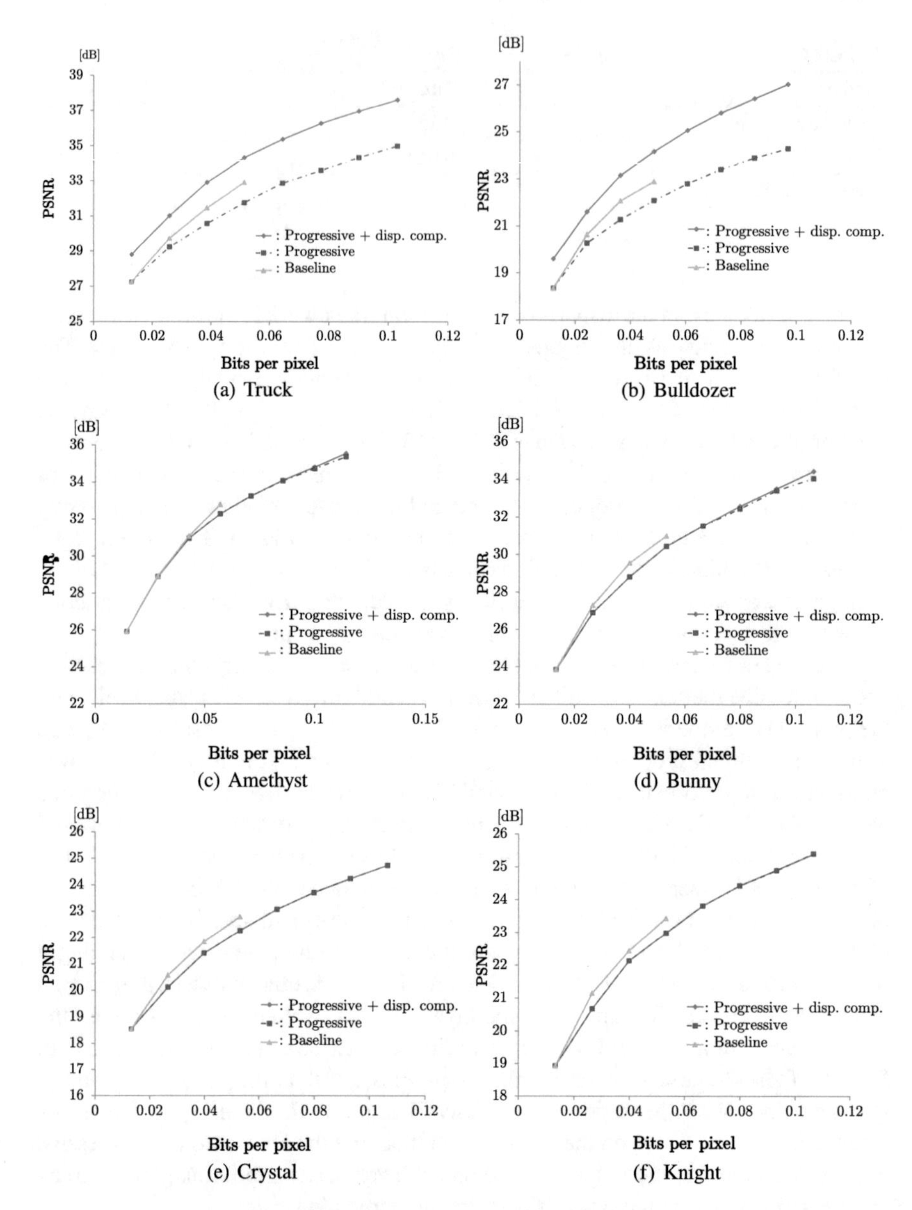

Fig. 7 R-D curves of the proposed method (Progressive + disp. comp.), the conventional progressive method (Progressive), and the conventional baseline method (Baseline)

Table 3 Selected disparity values in each layer

layer number	1	2	3	4	5	6	7	8
Truck	0.2	0.0	0.2	0.0	0.2	0.0	0.0	0.2
Bulldozer	0.2	0.2	0.8	0.2	0.2	0.2	0.5	−0.2
Amethyst	0.0	0.0	0.0	0.0	0.0	−0.2	0.0	0.0
Bunny	0.0	0.0	0.0	0.0	0.0	−0.2	0.0	0.0
Crystal	0.0	0.0	0.0	0.0	0.0	0.0	0.0	0.0
Knight	0.0	0.0	0.0	0.0	0.0	0.0	−0.5	0.0

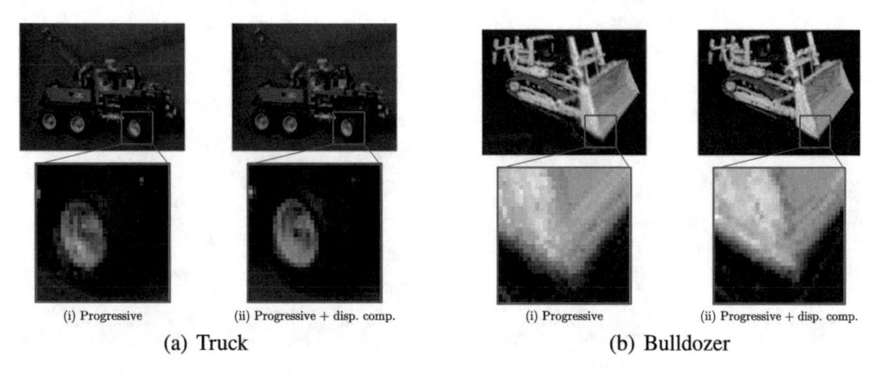

(a) Truck (b) Bulldozer

Fig. 8 Visual comparison of most top-left image ($N = 24$)

Table 4 Encoding time for Truck [s]

# of binary patterns	3	6	9	12
Baseline	16.9	106	876	7521
Progressive	16.8	32.0	49.0	64.1
Progressive + disp. comp.	263	514	774	1026

method improves rate-distortion performance while avoiding a computational complexity explosion.

Next, the proposed method was compared with the modern video coding standard H.265/HEVC. As general implementations of H.265/HEVC, the HEVC Test Model [5] was used with random access configuration and FFmpeg ver. 4.1. To apply these video codecs to light field datasets, images in the dataset in the row-major order were aligned and were regarded as a video sequence. The bitrate of the proposed method was calculated from the binary patterns, weights, and used disparity values, which are compressed by gzip ver. 1.9. As a reference, the baseline method was also compared with the above methods.

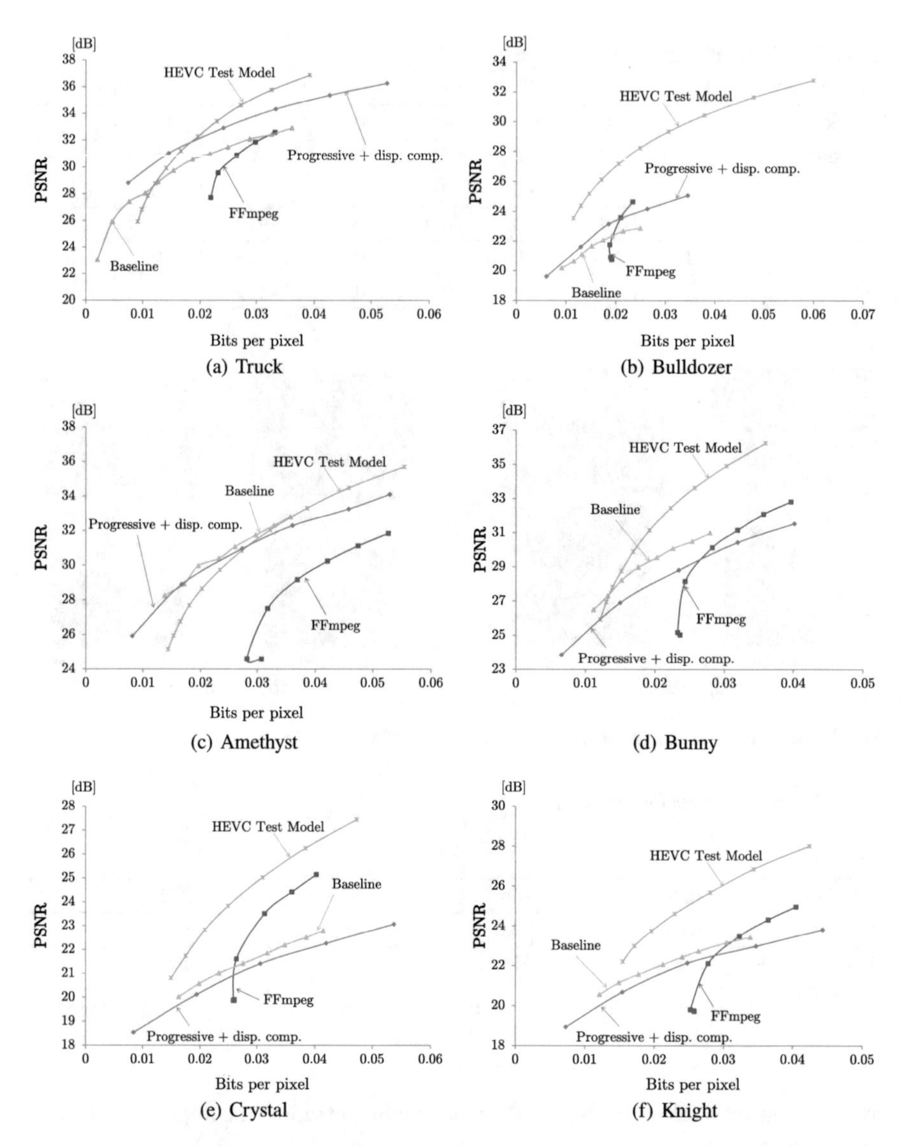

Fig. 9 Comparison with modern video coding standard H.265/HEVC

Figure 9 shows rate-distortion performances of three methods for six datasets. The proposed method shows better performance than that of the baseline method in truck dataset, and the performance can be more comparable to that of the HEVC Test Model. However, the proposed method still shows slightly inferior performance to the HEVC Test Model in the other datasets. As mentioned in previous section, the authors still believe that the performance of the proposed method is promising because it achieves superior or comparable performance to that of FFmpeg and the

HEVC Test Model, which has been optimized using enormous labor and time. The excellent performance of HEVC comes from the combination of many sophisticated coding techniques such as intra/inter prediction, transform coding, and arithmetic coding, where the optimal coding modes (e.g., prediction mode and block partition) are selected in accordance with the image content. Meanwhile, the method with weighted binary patterns is merely constructed on a very simple framework using weighted binary patterns with disparity compensation.

1.5 Real-World Data Circulation in Light Field Processing

This subsection discusses data circulation in light field processing. Light field processing can be considered to consist of data acquisition, analysis, and implementation in real world as shown in Fig. 10. For data acquisition, light field data is captured by using camera systems such as a multi-camera array system and light field camera. For data analysis, various light field processing technologies can be considered as a light field analysis such as compression, calculation of layer patterns for a compressive display, 3D object shape estimation, etc. For implementation, 3D image/video systems such as light field display and free-viewpoint image/video reproduction system can be realized by using the analyzed data. Our laboratory inclusively studies on light field capturing, analysis, and display systems; the contents of this section especially focus on light field compression which is classified as the analysis part in data circulation. One example of light field data circulation is light field streaming system (Fig. 10). A target 3D object is captured by using a light field camera; and then, captured light field data should be compressed to transmit it. 3D display systems such as the compressive display reproduce the 3D object using transmitted light field data. For the output data of the display systems, subjective evaluation can be obtained from viewers. By analyzing the result of evaluation, what is the important for 3D display systems would be found. Although what is important for 3D display system depends on the purpose of the system, each circulation stage could also be improved depending on the subjective evaluation; consequently, light field data circulation could be realized.

More specifically, data circulation in the study on light field coding can be considered as shown in Fig. 11. A room for improving compression schemes is first found by considering current performance of coding schemes and demands for practical systems. The performance of existing schemes such as rate-distortion characteristics, encoding/decoding times, and reconstructed images are obtained by experiments, where an input light field is encoded/decoded and data size, computational time for encoding/decoding, and visual quality are obtained. In the case of methods based on the modern video coding standards, these data can be obtained by decoding the bitstream, which is an output of the coding standards. Then, room for improving the scheme can be discussed based on the obtained data. For instance, degradation of decoded images can be confirmed by visual comparison and cause of the degradation could be found by carefully observing it. When the method of applying disparity com-

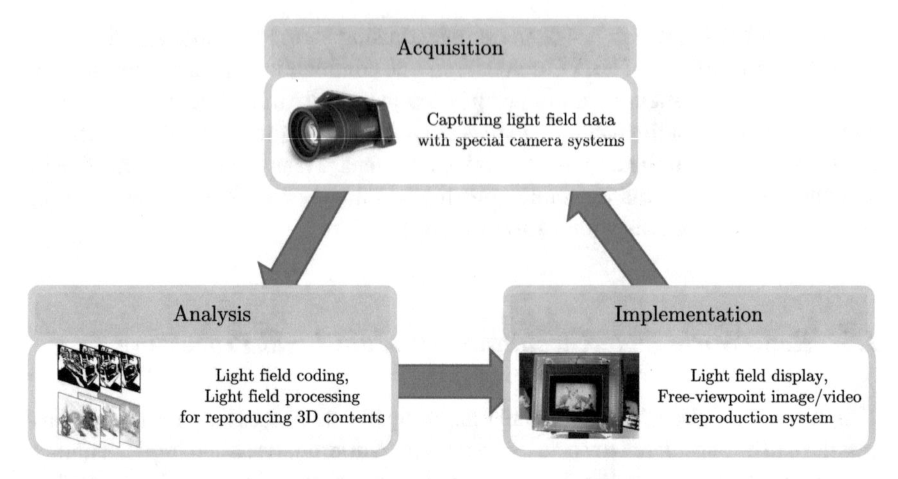

Fig. 10 Real-world data circulation in light field processing

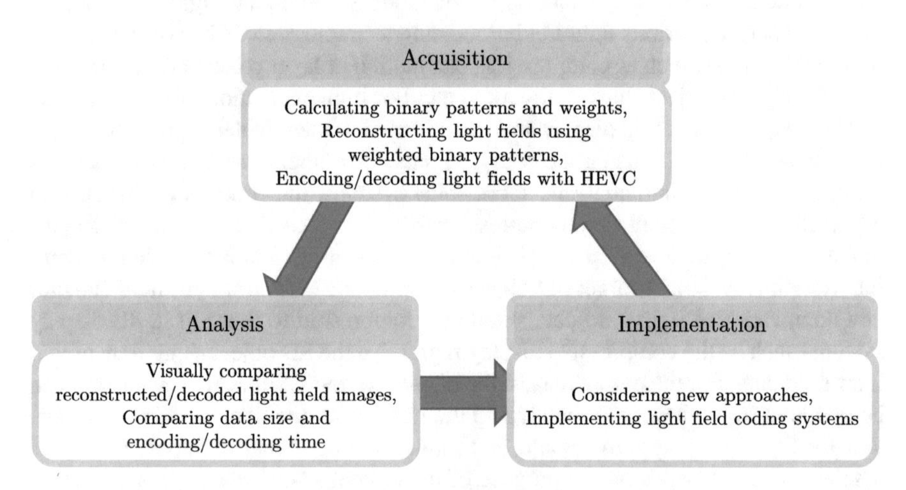

Fig. 11 Real-world data circulation in a study on efficient light field coding

pensation to the progressive coding was considered, large degradation of decoded images was able to be seen on a part having large disparities by comparing the decoded images with original one. Based on the result of the discussion, approaches for light field compression could be designed. After designing approaches, the approaches are actually implemented and experiments are conducted. Through the experiments, new data of the considered light field coding scheme can be obtained and the remaining issues and new approaches can be discussed based on the obtained data. This procedure could be considered as one of data circulation.

References

1. Adelson EH, Wang JYA (1992) Single lens stereo with a plenoptic camera. IEEE Trans Pattern Anal Mach Intell 14(2):99–106
2. Bossen F (2013) Common HM test conditions and software reference configurations
3. Computer Graphics Laboratory, Stanford University (2008) The (new) stanford light field archive. http://lightfield.stanford.edu/lfs.html
4. Flynn J, Neulander I, Philbin J, Snavely N (2016) Deepstereo: learning to predict new views from the world's imagery. 2016 IEEE conference on computer vision and pattern recognition (CVPR) pp 5515–5524
5. Fraunhofer Heinrich Hertz Institute (2018) High efficiency video coding (HEVC). https://hevc.hhi.fraunhofer.de
6. Fujii T, Kimoto T, Tanimoto M (1996) Ray space coding for 3D visual communication. In: Picture coding symposium (PCS), pp 447–451
7. Fujii T, Mori K, Takeda K, Mase K, Tanimoto M, Suenaga Y (2006) Multipoint measuring system for video and sound - 100-camera and microphone system. In 2006 IEEE international conference on multimedia and expo, pp 437–440. https://doi.org/10.1109/ICME.2006.262566
8. Fujii Laboratory, Nagoya University (2018) Light field compression project. http://www.fujii.nuee.nagoya-u.ac.jp/~takahasi/Research/LFCompression/index.html
9. Georgiev T, Lumsdaine A (2012) The multi-focus plenoptic camera. Digital Photography VIII 8299:69–79. https://doi.org/10.1117/12.908667
10. Gortler SJ, Grzeszczuk R, Szeliski R, Cohen MF (1996) The lumigraph. In: ACM SIGGRAPH, pp 43–54
11. Hawary F, Guillemot C, Thoreau D, Boisson G (2017) Scalable light field compression scheme using sparse reconstruction and restoration. 2017 IEEE international conference on image processing (ICIP) pp 3250–3254. https://hal.archives-ouvertes.fr/hal-01597536
12. Houben G, Fujita S, Takahashi K, Fujii T (2018) Fast and robust disparity estimation for noisy light fields. In: 2018 25th IEEE international conference on image processing (ICIP), pp 2610–2614. https://doi.org/10.1109/ICIP.2018.8451833
13. International Telecommunication Union Telecommunication Standardization Sector (2018) ITU-T Recommendation H.265 and ISO/IEC 23008-2: High efficiency video coding
14. Isechi K, Takahashi K, Fujii T (2020) [paper] disparity compensation framework for light-field coding using weighted binary patterns. ITE Trans Media Technol Appl 8(1):40–48. https://doi.org/10.3169/mta.8.40
15. Jiang X, Pendu ML, Farrugia RA, Guillemot C (2017) Light field compression with homography-based low-rank approximation. IEEE J Selected Topics Signal Process 11(7):1132–1145
16. Komatsu K, Isechi K, Takahashi K, Fujii T (2019) Light field coding using weighted binary images. IEICE Trans Inform Syst E102-D(11):2110–2119
17. Lanman D, Hirsch M, Kim Y, Raskar R (2010) Content-adaptive parallax barriers: Optimizing dual-layer 3D displays using low-rank light field factorization. ACM Trans Graph 29(6):163:1–163:10
18. Lee S, Jang C, Moon S, Cho J, Lee B (2016) Additive light field displays: realization of augmented reality with holographic optical elements. ACM Trans Graph 35(4):60:1–60:13
19. Levoy M, Hanrahan P (1996) Light field rendering. In: Proceedings of the 23rd annual conference on computer graphics and interactive techniques, ACM, New York, NY, USA, SIGGRAPH '96, pp 31–42
20. Li Y, Sjöström M, Olsson R, Jennehag U (2014) Efficient intra prediction scheme for light field image compression. In: 2014 IEEE international conference on acoustics, speech and signal processing (ICASSP), pp 539–543
21. Liu D, Wang L, Li L, Xiong Z, Wu F, Zeng W (2016) Pseudo-sequence-based light field image compression. In: 2016 IEEE international conference on multimedia expo workshops (ICMEW), pp 1–4

22. Lumsdaine A, Georgiev T (2009) The focused plenoptic camera. In: IEEE international conference on computational photography (ICCP), pp 1–8
23. Ng R, Levoy M, Brédif M, Duval G, Horowitz M, Hanrahan P (2005) Light field photography with a hand-held plenoptic camera. Comput Sci Techn Report CSTR 2(11):1–11
24. Saito T, Kobayashi Y, Takahashi K, Fujii T (2016) Displaying real-world light fields with stacked multiplicative layers: Requirement and data conversion for input multiview images. J Display Technol 12(11):1290–1300. https://doi.org/10.1109/JDT.2016.2594804
25. Shi L, Hassanieh H, Davis A, Katabi D, Durand F (2014) Light field reconstruction using sparsity in the continuous Fourier domain. ACM Trans Graph 34(1):12:1–12:13
26. Suzuki T, Takahashi K, Fujii T (2016) Disparity estimation from light fields using sheared EPI analysis. In: 2016 IEEE international conference on image processing (ICIP), pp 1444–1448
27. Suzuki T, Takahashi K, Fujii T (2017) Sheared EPI analysis for disparity estimation from light fields. IEICE Trans Inform Syst E100.D(9):1984–1993
28. Tech G, Chen Y, Müller K, Ohm JR, Vetro A, Wang YK (2016) Overview of the multiview and 3D extensions of high efficiency video coding. IEEE Trans Circuits Syst Video Technol 26(1):35–49
29. Vieira A, Duarte H, Perra C, Tavora L, Assuncao P (2015) Data formats for high efficiency coding of lytro-illum light fields. In: 2015 international conference on image processing theory, tools and applications (IPTA), pp 494–497. https://doi.org/10.1109/IPTA.2015.7367195
30. Wanner S, Goldluecke B (2014) Variational light field analysis for disparity estimation and super-resolution. IEEE Trans Pattern Analysis Mach Intell 36(3):606–619
31. Wetzstein G, Lanman D, Hirsch M, Raskar R (2012) Tensor displays: Compressive light field synthesis using multilayer displays with directional backlighting. ACM Trans Graph 31(4):80:1–80:11. https://doi.org/10.1145/2185520.2185576
32. Wilburn B, Joshi N, Vaish V, Talvala EV, Antunez E, Barth A, Adams A, Horowitz M, Levoy M (2005) High performance imaging using large camera arrays. ACM Trans Graph 24(3):765–776. https://doi.org/10.1145/1073204.1073259

Point Cloud Compression for 3D LiDAR Sensor

Chenxi Tu

Abstract Point cloud data from LiDAR sensors is currently the basis of most Level 4 autonomous driving systems, and its use is expanding into many other fields. The sharing and transmission of point cloud data from 3D LiDAR sensors has broad application prospects in areas such as accident investigation and V2V/V2X networks. Due to the huge volume of data involved, directly sharing and storing this data is expensive and difficult; however, making compression indispensable. Many previous studies have proposed methods of compressing point cloud data. Because of the sparseness and disorderly nature of this data, most of these methods involve arranging point clouds into a 2D format or into a tree structure and further coding them, while these converting methods usually lead to information loss. In my research, we propose a new formatting method to losslessly format point cloud. Variant approaches are then proposed to reduce spatial and temporal redundancy, which can obviously outperform previous methods. At the same time, my research is closely related to real-world data circulation (RWDC). It can advance RWDC by helping storing long-term data. And to those learning-based approach (Sects. 5 and 6), my compression method can be a RWDC itself.

1 Introduction

1.1 Background

A point cloud is a collection of points spread across a 3D space, which can be thought of as a sampling of the real world. Point cloud data collected by LiDAR sensors is currently being used for driving environment representation by many autonomous driving systems [7], including those operated by Google and Uber [9].

Meanwhile, shared and stored streaming point cloud data is likely an important component of accident investigation, future V2V (vehicle-to-vehicle) and V2X

C. Tu (✉)
Nagoya University, Longgang District, Shenzhen, China
e-mail: tu.chenxi@g.sp.m.is.nagoya-u.ac.jp

K. Takeda et al. (eds.), *Frontiers of Digital Transformation*,
https://doi.org/10.1007/978-981-15-1358-9_8

(vehicle-to-everything) systems [2] for autonomous driving. Sharing and storing point cloud from LiDAR is also meaningful for many real robotics applications, for example, remote control. Many robotic applications employ remote control systems based on camera sensors, such as [6]. A 3D LiDAR scanner generating point cloud of local environment could be a good alternative or supplement to video based control, like the case in [10]. Another possible application is multi-robot synergy. Researchers have used laser scanners for mapping tasks involving multiple robots [13], in which the data collected by the robot-mounted sensors must be transmitted using limited bandwidth [16].

Streaming point cloud data is a type of "big data"; however, one hour of point cloud data from the Velodyne HDL-64 sensor mentioned above can result in over 100GB of data, which is too large to realistically share or store using currently available technology. Thus, developing methods of compressing this data has become an indispensable task.

1.2 Point Cloud Compression Challenges

In order to compress point cloud data from LiDAR sensors, we need to eliminate both spatial and temporal redundancy. This poses several problems, as shown in Fig. 1.

The first and biggest problem is that it is difficult to directly reduce a point cloud's spatial redundancy because of its sparsity and disorder. Therefore, point clouds are usually converted into another format before spatial redundancy can be effectively reduced. Most compression approaches directly format point cloud data into a tree structure or 2D matrix, but all of these methods lead to information loss during the reformatting process. To avoid this, we need to develop a new approach which allows us to losslessly reformat point cloud data. However, with the new data format, a suitable approach for reducing spatial redundancy is also needed, which is another challenge.

Reducing temporal redundancy in streaming point cloud data is a more complex problem than reducing spatial redundancy, and performance is highly dependent on the format of the point cloud. Taking a portion of the frames as reference/key frames and then using these frames to predict the content of the enclosed frames is a popular

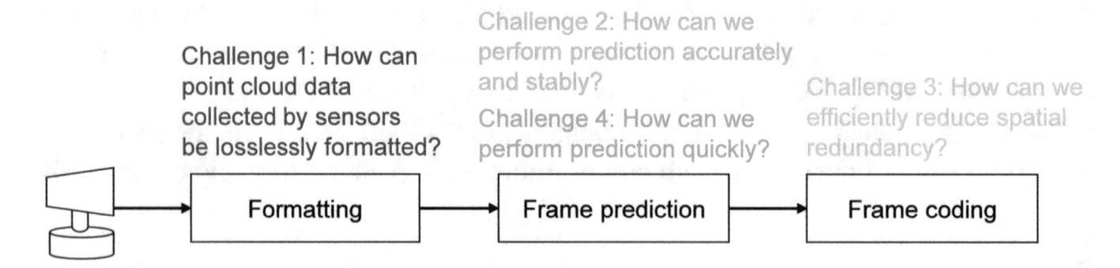

Fig. 1 Point cloud compression challenges

strategy for compressing streaming data. One example of this approach is MPEG compression, which uses motion compensation to predict video frames. In the case of streaming point cloud data, the key to reducing temporal redundancy is developing a method for efficiently predicting the content of the LiDAR frames bracketed between pairs of reference frames.

A good frame prediction method usually needs to be accurate, steady, and fast. However, accurate and steady prediction usually requires a huge amount of computation, which conflicts with fast processing. In this paper, we treat achieving accurate, steady prediction and achieving high processing speed as two separate problems.

1.3 Contribution

To tackle these challenges, four different compression methods are proposed. These methods are introduced in Sects. 3, 4, 5, and 6.

As discussed in Sect. 1.2, there are four big challenges to be overcome in order to efficiently compress point clouds from LiDAR; lossless reformatting of point cloud data, stable and accurate frame prediction, efficient reduction of spatial redundancy, and quick frame prediction. Section 3 introduces a method of losslessly reformatting point cloud data, which solves the first challenge. An existing image/video compression method is used to further compress the data, which is regarded as a baseline in following chapters. Section 4 targets the second challenge of stable and accurate frame prediction. In order to achieve compression with high accuracy and stability, a LiDAR simulation-based method is proposed. Section 5 focuses on the third challenge, reducing spatial redundancy efficiently, and proposes an RNN-based approach to compress a single point cloud frame. Section 6 concentrates on high-speed processing in order to tackle the fourth challenge of quick frame prediction. A U-net based method is proposed to compress streaming point cloud from LiDAR in real time.

2 Related Work

In order to effectively compress static point cloud data, we need to reduce its spatial redundancy. If we are compressing streaming point cloud data, we also need to reduce temporal redundancy.

Reducing spatial redundancy is important when compressing both static and streaming point cloud data. However, since point clouds are generally sparse and disorderly, they are difficult to compress in their original format. Therefore, before compression, point clouds are generally converted into a different format, and then quantized and encoded, in order to reduce spatial redundancy. Many studies have been published proposing various approaches for formatting and compressing point cloud data, as shown in Fig. 2.

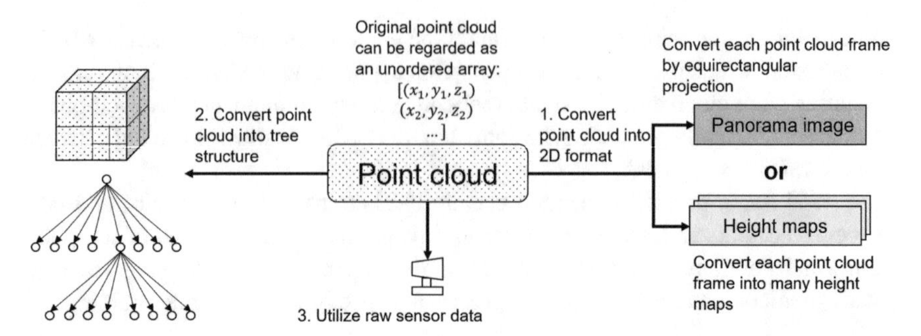

Fig. 2 Many methods of compressing sparse and disorderedly point cloud data have been proposed. Most of these strategies fall into one of the following categories: (1) converting 3D point cloud data into a 2D format, (2) converting 3D point cloud data into a tree structure, (3) utilizing raw sensor data to avoid directly processing the point cloud data

Most previous point cloud compression methods have been designed to compress a single, static point cloud, and thus have only focused on reducing spatial redundancy, making these approaches unsuitable for the compression of streaming data. The streaming point cloud compression methods which have been proposed, such as [4, 15], attempt to reduce both spatial and temporal redundancy.

The key to reducing temporal redundancy is the efficient utilization of reference frames to predict the remaining frames. One classic strategy is differential encoding, which calculates and transmits the residual between adjacent signals/frames. Kammer et al. [4] propose an octree-based differential encoding algorithm for compressing point cloud streams. They used an octree-based method to represent the spatial location of each point, and a double-tree structure to calculate the difference (exclusive or) between the octrees of adjacent frames, allowing the reduction of time series redundancy for streaming. Kammer's method can be used to control information loss quantitatively, and allows the real-time compression of streaming point cloud data. Since the open code for this method is available to other researchers via the Point Cloud Library, it has become a popular method for the compression of streaming point cloud data.

Another strategy for reducing temporal redundancy is to calculate motion vectors for consecutive frames, a method widely used in the compression of video and 3D meshes. Thanou et al. [15] proposed a method of streaming point cloud data compression which includes a spatial motion estimation module. Octrees are used to divide 3D point clouds into many occupied voxels, while the points in each voxel (or "leaf") are represented by a weighted, undirected graph.

3 Data Formatting and Image/Video-Based Compression Methods

As mentioned before, since point cloud data are sparse and disorderly, before reducing spatial redundancy the data are generally converted into another format. While directly converting point cloud data into a 2D format or tree structure results in information loss, this section suggests an approach which utilizes the LiDAR packet data, allowing the lossless conversion of a point cloud's data points into a 2D matrix format, which can solve the challenge 1 mentioned in Fig. 1.

LiDAR sensors detect their surroundings by emitting pulses of laser energy and detecting their reflections. Each point in point cloud can be represented by a set of $x-$, $y-$, and z-coordinates, which identifies the point's location in a 3D space. In contrast, raw LiDAR packet data (R) represent each point using a distance value, a rotation angle, and a laser ID. Rotation angle here refers to the yaw angle of a LiDAR at each emission. Note that a particular laser's emissions occur at a particular yaw angle once per rotation, without taking calibration into consideration. Laser ID here represents pitch angle information. In LiDAR systems, every laser sensor is fixed at a specific location and angle, so that if the laser ID is known we can easily determine the pitch angle of the beam. In other words, raw packet data can be roughly considered as a kind of polar-coordinate-like representation of a 3D point cloud. After a calibration process ($f(R) = P$), which uses a calibration file to correct the yaw angle, distance, and starting locations, raw packet data R can be converted into a point cloud P in real time.

Because raw LiDAR packet data has a natural 2D format, we can easily and losslessly arrange it into a 2D matrix by making each row correspond to one laser ID, each column correspond to once emission, and the value of each pixel represents distance information. Figure 3 shows an example of an image created using this type of raw packet data. Note that, on account of the internal structure of the LiDAR system, 2D matrices of raw LiDAR packet data are irregular, i.e., the data is not directly spatially correlated and cannot be understood intuitively.

After converting streaming point cloud data from a LiDAR system into a video-like format, utilizing an image/video compression method to further compress the data is a logical next step. We explore the use of existing image/video compression

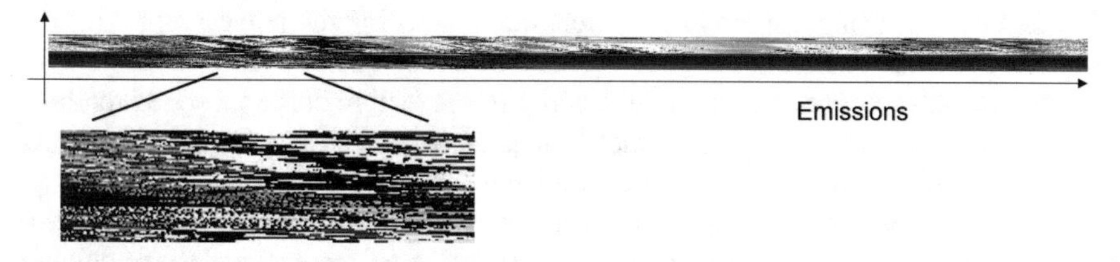

Fig. 3 Visualization of raw LiDAR packet data in an image-like format. A pixel's grayscale value from black to white represents a distance from 0 to 13,000 cm. Noted that without calibration, we could not understand this raw data in the way one understands a depth map

methods (JPEG and MPEG) with pre-processing to further compress 2D formatted LiDAR data. These methods are also regarded as baselines in the following Sections.

4 Compression Based on LiDAR Simulation and Motion Analysis

The performance of streaming data compression methods always depends on their efficiency at reducing temporal redundancy. Section 3 proposed the use of existing video compression methods to compress streaming LiDAR point cloud data. In contrast to video data, however, the pixels in adjacent frames of 2D formatted LiDAR data not only translate, but their values (usually representing distance) also change. In addition, adjacent pixels do not always have similar motion. All of these phenomena make it difficult to directly use the pixels of reference frames to predict the content of nearby frames, as is done during video compression.

To reduce temporal redundancy efficiently, we need to predict frames accurately and stably, as challenge 2 in Fig. 1 mentions. For this purpose, this section proposes a frame prediction method to reduce temporal redundancy in streaming LiDAR packet data by simulating the operation of a LiDAR system, and by utilizing LiDAR motion information. Furthermore, based on this analysis of the motion of the LiDAR unit, a dynamic approach to reference frame distribution is proposed, which further improves compression performance. This LiDAR simulation and motion analysis-based compression method was first proposed by us at the 2017 IEEE Intelligent Vehicles Symposium (IV2017) [19], and we developed it further in a subsequent study [20].

The proposed method can be divided into three components: (a) motion analysis, (b) prediction, and (c) encoding, as shown in Fig. 4. The motion analysis module selects reference frames from the data, the prediction module reduces temporal redundancy, and the encoder reduces spatial redundancy.

Motion analysis (a) is the first step in the compression process, and is composed of two sub-processes: motion estimation and sequencing. Motion estimation here means obtaining the rotation (yaw, pitch, roll) and translation (x, y, z) of the LiDAR data in each frame. The proposed method uses Simultaneous Localization and Mapping (SLAM) based on the Normal Distributions Transform (NDT) [5, 14] to obtain this information. By using motion information, the sequencing module can optimize the number and location of the reference frames. The use of a motion analysis module for streaming data compression is a unique and original feature of the proposed method. Other streaming data compression methods utilize a prediction + encoder structure, but the reference frames are usually chosen at a constant interval.

Prediction (b) is the second step of the proposed method. In the following sections, we call the reference frames 'I-frames' (intra-coded frames) and the remaining frames, which are predicted using the reference frames, 'B-frames' (bi-directionally predicted frames), which are conventional designations used in the video compres-

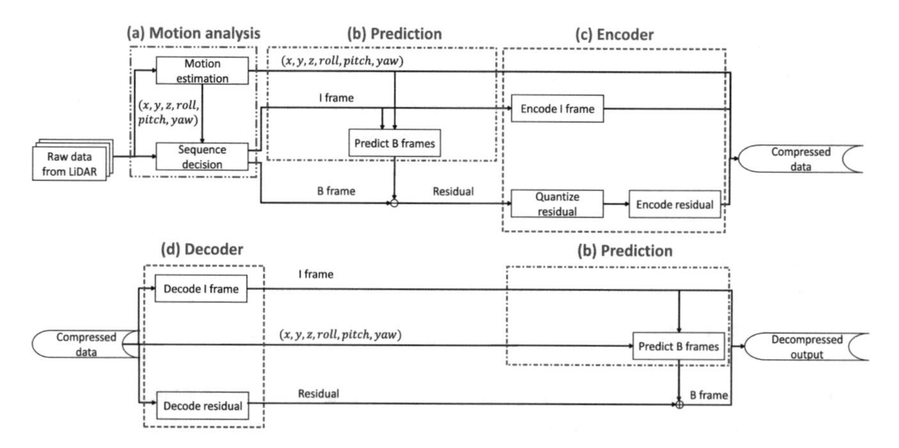

Fig. 4 Flowcharts illustrating the proposed LiDAR simulation-based method. The upper flowchart shows the compression process, while the lower one shows decompression

sion field. The prediction module attempts to predict the B-frames using data from the I-frames and the motion estimation results. The residuals between the predicted B-frames and the true B-frames are then calculated.

Encoding (c) is the last step of the compression process, in which the I-frames are compressed losslessly, while the residuals of the B-frames are quantized and coded.

The decoding process (d) can be thought of as the inverse of the process conducted during encoding.

5 RNN-Based Single Frame Compression

In this section, we focus on reducing spatial redundancy more efficiently (i.e., the challenge 3 shown in Fig. 1) and target at static point cloud compression, which is a component of streaming point cloud compression.

Section 3 is introduced to arrange LiDAR packet data into a 2D matrix naturally and losslessly to convert into point cloud using calibrations. The challenge is that, on account of the structure of the LiDAR, 2D matrices of raw LiDAR packet data are irregular, i.e., the data are not directly spatially correlated and cannot be understood intuitively. Some of these raw LiDAR packet data compression methods sacrifice accuracy and reduce the number of bits used by each "pixel" [23], or use existing image compression methods to compress the 2D matrix, as introduced in Sects. 3 and 4. Considering the irregularity of image-like raw packet data, these compression methods are not suitable.

Deep learning has already achieved state-of-the-art results in many fields, and data compression may be a task which deep learning is good at. In this section, a compression method which uses a recurrent neural network with residual block structures is proposed to compress one frame of point cloud data progressively.

5.1 Method

The overall approach can be divided into three parts. First, we convert one frame of raw packet data into a 2D matrix $\boldsymbol{R}$ plus a few extra bits, as introduced in Sect. 3. Second, we conduct pre-processing to normalize the data before sending it along the network. Third, a recurrent neural network composed of an encoder, a binarizer, and a decoder is used for data compression. In contrast to a related study [17], a new type of decoder network is proposed which uses residual blocks to improve the decompression performance.

As shown in Fig. 5, the compression network is composed of an encoding network E, a binarizer B, and a decoding network D, where D and E contain recurrent network components. The original 2D matrix $\boldsymbol{R}$ is first sent into the encoder, and then the binarizer transforms the encoder's output into a binary file which can be stored as the compression output and transmitted to the decoder. The decoder tries to reconstruct the original input using the received binary file. This process represents one iteration. The calculated residual between the original input data and the reconstructed data becomes the input of the next iteration. Repeating this, after one more iteration, we obtain more bits of compression output while the decompressed data can be more accurate.

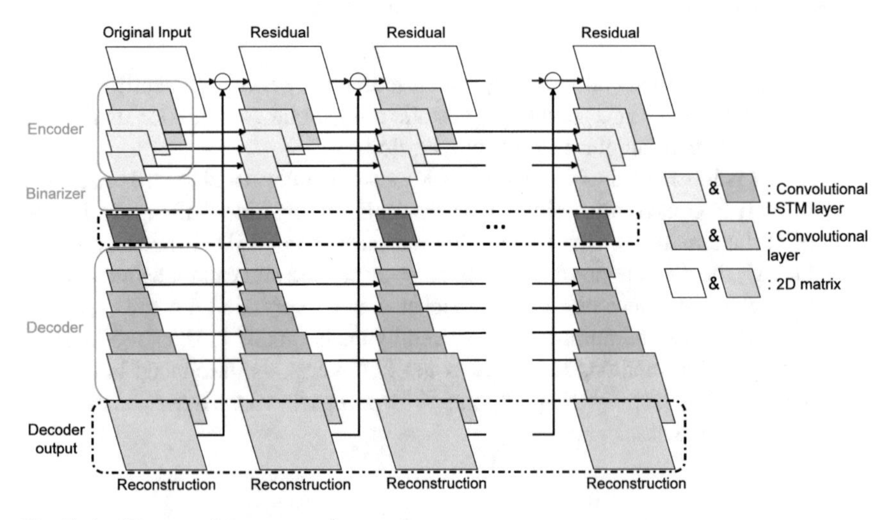

Fig. 5 Architecture of the proposed network

We can compactly represent a single iteration of this network as follows:

$$b_t = B(E_t(r_{t-1})), \quad \hat{x}_t = D_t(b_t) + \hat{x}_{t-1}, \tag{1}$$

$$r_0 = x, \quad \hat{x}_0 = 0 \tag{2}$$

Subscript t here represents the iteration, for example, D_t and E_t represent the decoder and encoder with their states at iteration t, respectively. x represents the original input, $\hat{x}_t$ the progressive reconstruction of the original input, r_t the residual between x and the reconstruction $\hat{x}_t$, $b_t \in \{-1, 1\}^m$ the m bits binarized stream from B. After k iterations, compression output totally needs $m \times k$ bits.

During training, the average residuals generated at each iteration is calculated as loss for the network:

$$\frac{1}{t} \sum_t |r_t| \tag{3}$$

Input x is a $32 \times 32 \times 1$ array. During training, a patch this size is randomly sampled from each training data, and during compression the height and the width of the input were made divisible by 32 through padding. After the encoder and the binarizer, the input is reduced to a $2 \times 2 \times 32$ binary representation per iteration, which leads to adding $1/8$ bit per point (bpp) for compression output after each iteration. By using more iterations, bits per point, in other words, volume needed by compression output, will increase linearly. At the same time, decompression can be more accurate.

During compression, data passes through the encoder, binarizer, and decoder, but during decompression, only the decoder is needed.

5.2 Evaluation

To train the network, we use 33,134 frames as training data, consisting 1 Hz sampling of driving data from 11 areas of Japan. All the point cloud data came from Velodyne HDL-32E sensor and include various situations like urban road, country road, forest park, etc. Test data come from a 32 min driving data in Akagi, Gunma, which is not one of the areas included in the training data.

In the evaluation, we compare the proposed method with JPEG image compression based approach [18] and the generally used octree compression. By taking advantage of the feature extraction and context analysis capability of RNN with convolutional layers, the proposed method can tune the compression rate with decompression error and greatly outperform previous image compression based approach and octree compression, which is constant to our hypothesis.

6 Real-Time Streaming Point Cloud Compression Using U-Net

Now, we return to the topic of streaming point cloud compression. In this section, in contrast to Section 4, we pay more attention to processing speed in order to address the challenge of quick frame prediction, mentioned in Sect. 1.2. To achieve real-time processing, a method of streaming point cloud compression using U-net is proposed.

6.1 Method

The overall structure of the proposed U-net-based compression method is very similar to the LiDAR simulation-based method introduced in Sect. 4. In contrast to the LiDAR simulation-based method, however, the U-net-based method does not rely on LiDAR motion information, so a motion analysis module is not needed. Without motion information to assist with segmentation, the number of B-frames between each pair of I-frames is fixed at a constant value by parameter n. The prediction part, which uses I-frames to predict the B-frame for reducing temporal redundancy, is now performed by a U-net-based network.

Inspired by Jiang et al.'s work [3], given two I-frames, I_0 and I_1, which are 2D matrices of reformatted LiDAR packet data, the proposed method uses two U-nets to infer the enclosed B-frames at time t (B_t). Here $t = 1/n, 2/n, \ldots (n-1)/n$.

Utilizing optical flow to interpolate 2D frames is a popular and efficient strategy, so the proposed method also uses this approach. To interpolate the B-frames between I_0 and I_1, optical flows $F_{0\to1}$ (from I_0 to I_1) and $F_{1\to0}$ (from I_1 to I_0) should be calculated first. Since U-net has proven its ability to calculate optical flow in many studies [11, 12], the proposed method also uses U-net to obtain $F_{0\to1}$ and $F_{1\to0}$, as shown in the blue box in Fig. 6. We call this U-net a "flow computation network."

Basically, given $F_{0\to1}$ and $F_{1\to0}$, we can linearly approximate $\hat{F}_{t\to1}$ and $\hat{F}_{t\to0}$ as follows:

$$\begin{aligned} \hat{F}_{t\to1} &= (1-t)F_{0\to1} \quad \text{or} \quad \hat{F}_{t\to1} = -(1-t)F_{1\to0} \\ \hat{F}_{t\to0} &= -tF_{0\to1} \quad \text{or} \quad \hat{F}_{t\to0} = tF_{1\to0}, \end{aligned} \tag{4}$$

Since, it is better to combine the bi-directional optical flows of $F_{0\to1}$ and $F_{1\to0}$ to calculate $F_{t\to1}$ and $F_{t\to0}$, we assume $F_{0\to1} = -F_{1\to0}$, and rewrite Eq. (4) as follows:

$$\begin{aligned} \hat{F}_{t\to0} &= -(1-t)tF_{0\to1} + t^2F_{1\to0} \\ \hat{F}_{t\to1} &= (1-t)^2F_{0\to1} - t(1-t)F_{1\to0}. \end{aligned} \tag{5}$$

Linearly approximating $F_{t\to1}$ and $F_{t\to0}$ is not sufficient to obtain the desired results; however, following Jiang et al.'s work [3], another U-net is used to refine $\hat{F}_{t\to1}$

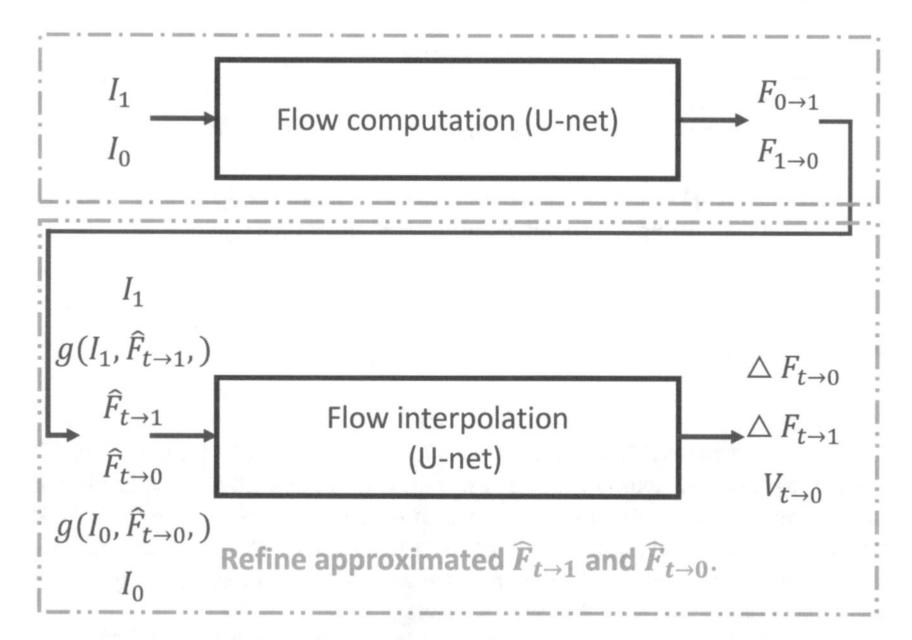

Fig. 6 Overview of entire frame interpolation network

and $\hat{F}_{t\to0}$, as shown in the red box in Fig. 6. We call this U-net a "flow interpolation network." The inputs of the interpolation network are I_0, I_1, $\hat{F}_{t\to1}$, $\hat{F}_{t\to0}$, $g(I_0, \hat{F}_{t\to0})$ and $g(I_1, \hat{F}_{t\to1})$. Here $g(\cdot, \cdot)$ is a *backward warping* function, which is implemented using bi-linear interpolation [8, 24]. In other words, we can approximate B_t using $g(I_0, \hat{F}_{t\to0})$ or $g(I_1, \hat{F}_{t\to1})$.

The outputs of the interpolation network are $\Delta F_{t\to0}$, $\Delta F_{t\to1}$ and $V_{t\leftarrow0}$. $\Delta F_{t\to0}$ and $\Delta F_{t\to1}$ are intermediate optical flow residuals. According to Jiang et al. [3], it is better to predict these residuals than to directly output the refined optical flow:

$$F_{t\to0} = \hat{F}_{t\to0} + \Delta F_{t\to0}$$
$$F_{t\to1} = \hat{F}_{t\to1} + \Delta F_{t\to1} \tag{6}$$

$V_{t\leftarrow0}$ is a weight matrix used to combine $g(I_0, F_{t\to0})$ and $g(I_1, F_{t\to1})$. As mentioned earlier, in contrast to the pixels in images, the 2D packet data pixels represent distances, which means their values will change in dynamic scenarios. Thus, optical flow alone is not enough to interpolate the enclosed B-frames; pixel values also need to be approximated from both $g(I_0, F_{t\to0})$ and $g(I_1, F_{t\to1})$. Assuming that a pixel's value changes linearly between two I-frames, we restrict the elements of $V_{t\leftarrow0}$ to a range from 0 to 1, resulting in

$$V_{t\leftarrow0} = 1 - V_{t\leftarrow1}. \tag{7}$$

Finally, we can obtain the interpolated B-frame $\hat{B}_t$:

$$\hat{B}_t = \frac{1}{Z} \odot \big((1-t)V_{t\leftarrow 0} \odot g(I_0, F_{t\rightarrow 0}) + tV_{t\leftarrow 1} \odot g(I_1, F_{t\rightarrow 1})\big)$$

where $Z = (1-t)V_{t\rightarrow 0} + tV_{t\rightarrow 1}$ is a normalization factor and $\odot$ denotes element-wise multiplication, implying content-aware weighting of the input.

6.2 Evaluation

For evaluation, the proposed U-net-based method is compared with other streaming point cloud compression methods, such as octree [4], MPEG [18], and LiDAR simulation-based methods [19, 21]. The proposed U-net-based compression method outperformed all of the other methods, except in some scenarios where the LiDAR simulation-based method achieved better performance. However, the LiDAR simulation-based method lacks the proposed method's real-time capability, which gives the U-net-based method a major advantage.

In the experiment, the original point cloud from LiDAR cost 244Mb/s. The proposed U-net-based method can compress them into 3.8 Mb/s, which is 1/64 of original data, with average 2 cm SNNRMSE as shown in Fig 7 and could even be affordable for nowadays 4G network.

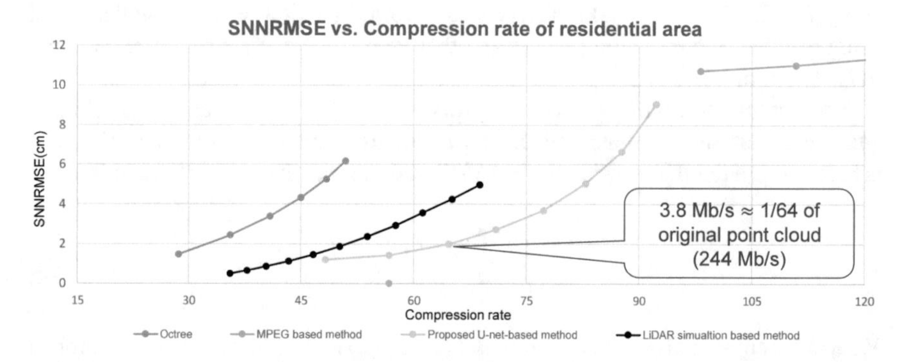

Fig. 7 SNNRMSE versus Compression rate for proposed U-net-based streaming point cloud compression method, in comparison with other compression methods. Symmetric Nearest Neighbor Root Mean Squared Error (SNNRMSE) is to measure loss after decompression

7 Relationship with Real-World Data Circulation

The concept of real-world data circulation (RWDC) is the transfer of data from its acquisition to its analysis, and in turn, to its implementation, which is a key process for the success of a commercial application. This section introduces the concept of RWDC and discusses the relationship between the presented research and RWDC from two aspects; how this research advances RWDC, and how itself is an example of an RWDC.

7.1 Definition of RWDC

Successful manufacturing requires that "real-world data" about the end users' status and expectations be continuously collected and applied to products and services [1]. Real-world data circulation (RWDC) refers to such a data-driven system. An RWDC usually consists of three parts: Acquiring real-world data, analyzing real-world data, and utilizing real-world data. One typical example of RWDC is an autonomous driving system, which collects driving data. By analyzing various types of driving data, we can further improve the safety and comfort of autonomous driving systems for their users. Another example is the development of pharmaceuticals, during which genomic information and medical histories are collected from patients. Omics analysis using this data can lead to the discovery of new drugs, which can then be used to treat patients.

7.2 Point Cloud Compression and RWDC

Compression technology is important for data storage and transmission, which can be thought of as a bridge between acquiring real-world data and analyzing it, which are the first two steps of RWDC. Effective compression methods can greatly improve the efficiency of RWDC loops by allowing the sharing of more data. At the same time, point cloud data compression has broad commercial applications in autonomous driving, augmented reality/virtual reality (AR/VR), in which a point cloud can be regarded as part of a 3D mesh, etc. All of the RWDC in these domains which involve point clouds are also likely to involve a compression method.

Let us take autonomous driving as a typical field for the application of RWDC. To develop a safe and comfortable autonomous driving system, big business's self-driving vehicles have driven tens of million kilometers. During this process, all the sensor data, algorithm output, and manual interventions need to be recorded to find out potential shortcomings or even to investigate accidents. Through these analysis, development department can improve the system, as shown in Fig. 8. An extreme case

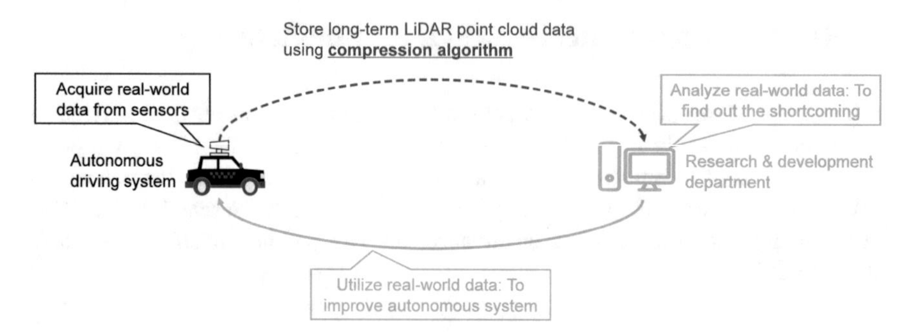

Fig. 8 RWDC system for autonomous driving using LiDAR point cloud data. Streaming point cloud compression technology, which can be regarded as a bridge, allows the storage and transmission of long-term real-world data before analysis

is the Uber's self-driving accident [22].[1] After that fatal crash, recorded sensor data and system log helped engineers to find out what happened and to further prevent it.

In order to make its data circulation process work, various types of data need to be stored and transmitted (i.e., the red dotted line in Fig. 8), and point clouds are no exception. As one of the most important types of sensor data used by autonomous driving systems, point clouds are widely used for localization and object detection.

To store long-term point cloud during driving, huge volume is needed without compression. Take Velodyne HDL-64 sensor as an example, directly storing streaming point cloud needs 285 Mb/s (i.e., over 128 GB/h). It is a very huge burden for storage device, considering about there are still other sensors, like cameras, data need to be stored. The proposed U-net-based method, which was introduced in Sect. 6, can compress data at 3.8 Mb/s, which means saving 98.6% volume. And as mentioned in Sect. 6, its computational cost is low enough for real-time processing. Even if not necessary, real-time compression can greatly help self-driving vehicles to store longer period data. Here, we use b to denote bitrate needed before compression (285 Mb/s in our case), $b' \ll b$ to denote bitrate needed after compression (3.8 Mb/s in our case). In an online system, if the algorithm needs nt time ($n \geq 1$) to compress t time's data, to store t time's data locally, we totally need $v(t)$ volume:

$$v(t) = bt + b'\frac{t}{n} - b\frac{t}{n} = \left(b - \frac{b}{n} + \frac{b'}{n}\right)t = \left[b - \frac{(b - b')}{n}\right]t \tag{8}$$

The smaller n is, the less volume is needed for the disk in a local self-driving car. In other words, we can store longer term data without changing disk. When $n = 1$, which means real-time compression, we can obtain the smallest $v(t)$ and store the longest data.

[1]On the night of March 18, 2018, an autonomous car operated by Uber—and with an emergency backup driver behind the wheel—struck and killed a woman on a street in Tempe, Ariz. It was believed to be the first pedestrian death associated with self-driving technology.

7.3 Proposed Method as RWDC

The investigation of streaming point cloud compression not only helps to facilitate RWDC by improving the efficiency of data transmission, but also the learning-based compression methods introduced in Sects. 5 and 6 are themselves a kind of RWDC system. The RNN and U-net-based methods require real-world data for training. Then, by analyzing large amounts of real-world data, the algorithms can extract efficient spatial and temporal high-level features of 2D formatted LiDAR point cloud data, allowing us to discard useless residual data. With the help of these compression algorithms, we can more easily collect, store, and transmit point cloud data. Theoretically speaking, the more data we have, the better learning network we can train, resulting in more efficient compression methods. Thus, we have the typical three-step process (acquiring real-world data, analyzing real-world data, and utilizing real-world data) of an RWDC system.

8 Expectation for Future RWDC

With the development of AI-related technology, as a kind of productive means, real-world data are becoming more and more important. We can expect that more value could be explored with the help of RWDC in the near further. One promising area is customized service. Due to the requirement of huge manpower from professionals, customized service is always a luxury in many domains. RWDC with individualized data has a potential to make customized service affordable to by normal families. An attractive field to me is garment customization. By analyzing body information collected by depth-camera or LiDAR, maybe we can generate customized clothes for customers. Medical treatment is another area which RWDC may be good at. Treatment of some difficult diseases, like cancers, still relies on doctors' experience nowadays. By analyzing real-world cases, maybe we can help doctors to decide the treatment plan.

I believe RWDC can contribute a lot for all human beings in the near future.

References

1. Anon. Graduate program for real-world data circulation leaders. http://www.rwdc.is.nagoya-u.ac.jp/
2. Harding J, Powell G, Yoon R, Fikentscher J, Doyle C, Sade D, Lukuc M, Simons J, Wang J (2014) Vehicle-to-vehicle communications: readiness of V2V technology for application. Technical report
3. Jiang H, Sun D, Jampani V, Yang MH, Learned-Miller E, Kautz J (2018) Super slomo: high quality estimation of multiple intermediate frames for video interpolation. In: Proceedings of 2018 IEEE conference on computer vision and pattern recognition (CVPR), pp 9000–9008

4. Kammerl J, Blodow N, Rusu RB, Gedikli S, Beetz M, Steinbach E (2012) Real-time compression of point cloud streams. In: Proceedings of 2012 IEEE international conference on robotics and automation (ICRA), pp 778–785
5. Kato S, Takeuchi E, Ishiguro Y, Ninomiya Y, Takeda K, Hamada T (2015) An open approach to autonomous vehicles. IEEE Micro 35(6):60–68
6. Kim JW, Choi BD, Park SH, Kim KK, Ko SJ (2002) Remote control system using real-time MPEG-4 streaming technology for mobile robot. In: Digest of technical papers. International conference on consumer electronics, pp 200–201
7. Levinson J, Askeland J, Becker J, Dolson J, Held D, Kammel S, Kolter JZ, Langer D, Pink O, Pratt V, et al. (2011) Towards fully autonomous driving: systems and algorithms. In: Proceedings of 2011 IEEE intelligent vehicles symposium (IV), pp 163–168
8. Liu Z, Yeh RA, Tang X, Liu Y, Agarwala A (2017) Video frame synthesis using deep voxel flow. In: Proceedings of 2017 IEEE international conference on computer vision (ICCV), pp 4463–4471
9. Margulis C, Goulding C (2017) Waymo vs. Uber may be the next Edison vs. Westinghouse. J Pat Trademark Off Soc 99:500
10. Nagatani K, Kiribayashi S, Okada Y, Otake K, Yoshida K, Tadokoro S, Nishimura T, Yoshida T, Koyanagi E, Fukushima M et al (2013) Emergency response to the nuclear accident at the Fukushima Daiichi Nuclear Power Plants using mobile rescue robots. J Field Robot 30(1):44–63
11. Niklaus S, Liu F (2018) Context-aware synthesis for video frame interpolation. In: Proceedings of 2018 IEEE conference on computer vision and pattern recognition (CVPR), pp 1701–1710
12. Niklaus S, Mai L, Liu F (2017) Video frame interpolation via adaptive separable convolution. In: Proceedings 2017 IEEE international conference on computer vision (CVPR), pp 261–270
13. Rooker MN, Birk A (2007) Multi-robot exploration under the constraints of wireless networking. Control Eng Pract 15(4):435–445
14. Takeuchi E, Tsubouchi T (2006) A 3-D scan matching using improved 3-D normal distributions transform for mobile robotic mapping. In: Proceedings of 2016 IEEE intelligent robots and systems (IROS), pp 3068–3073
15. Thanou D, Chou PA, Frossard P (2016) Graph-based compression of dynamic 3D point cloud sequences. IEEE Trans Image Process 25(4):1765–1778
16. Thrun S, Burgard W, Fox D (2000) A real-time algorithm for mobile robot mapping with applications to multi-robot and 3D mapping. In: Proceedings of 2000 IEEE international conference on robotics and automation (ICRA), vol 1, pp 321–328
17. Toderici G, Vincent D, Johnston N, Hwang SJ, Minnen D, Shor J, Covell M (2017) Full resolution image compression with recurrent neural networks. In: Proceedings of 2017 IEEE conference on computer vision and pattern recognition (CVPR), pp 5435–5443
18. Tu C, Takeuchi E, Miyajima C, Takeda K (2016) Compressing continuous point cloud data using image compression methods. In: Proceedings of 2016 IEEE international conference on intelligent transportation systems (ITSC), pp 1712–1719
19. Tu C, Takeuchi E, Miyajima C, Takeda K (2017) Continuous point cloud data compression using SLAM based prediction. In: Proceedings of 2017 IEEE intelligent vehicles symposium (IV), pp 1744–1751
20. Tu C, Takeuchi E, Carballo A, Miyajima C, Takeda K (2019a) Motion analysis and performance improved method for 3D LiDAR sensor data compression. IEEE Trans Intell Transp Syst 1–14. https://doi.org/10.1109/TITS.2019.2956066
21. Tu C, Takeuchi E, Carballo A, Takeda K (2019b) Real-time streaming point cloud compression for 3d lidar sensor using u-net. IEEE Access 7:113616–113625
22. Wakabayashi D (2018) Self-driving uber car kills pedestrian in Arizona, where robots roam. https://www.nytimes.com/2018/03/19/technology/uber-driverless-fatality.html
23. Yin H, Berger C (2017) Mastering data complexity for autonomous driving with adaptive point clouds for urban environments. In: Proceedings of 2017 IEEE intelligent vehicles symposium (IV), pp 1364–1371
24. Zhou T, Tulsiani S, Sun W, Malik J, Efros AA (2016) View synthesis by appearance flow. In: Proceedings of 2016 European conference on computer vision (ECCV), pp 286–301

Integrated Planner for Autonomous Driving in Urban Environments Including Driving Intention Estimation

Hatem Darweesh

Abstract Thousands are killed every day in traffic accidents, and drivers are mostly to blame. Autonomous driving technology is the ultimate technological solution to this problem. There are still many unresolved problems with autonomous driving technology, such as navigating complex traffic situations. One of the reasons is detecting other drivers' intentions. Planning, which determines the movement of autonomous vehicles, is the cornerstone of autonomous agent navigation. Planning applications consist of multiple modules with different interfaces. Another challenge is the lack of open-source planning projects that allow cooperation between development teams globally. In this chapter, I will introduce two approaches to the planning problem. The first is developing of an `open-source`, `integrated` planner for autonomous navigation called `Open Planner`. It is composed of a global path planner, intention predictor, local path planner, and behavior planner. The second is a novel technique for estimating the intention and trajectory probabilities of surrounding vehicles, which enables long-term planning and reliable decision-making. Evaluation was achieved using simulation and field experimentation.

1 Introduction

There are several causes for the high number of traffic fatalities and injuries; rapid urbanization, low safety standards, insufficient law enforcement, people driving while distracted, fatigued, or under the influence of drugs or alcohol, speeding and failure to wear seat belts or helmets. Drivers are responsible for most of these accidents, as opposed to equipment failure or road hazards.

Many researchers are trying to tackle this problem by eliminating the human factor, using state-of-the art technology to replace human drivers with a collection of sensors, hardware, and software, in order to achieve automated driving. Self-driving vehicles (autonomous vehicles) are close to becoming a reality. Interest in this field was sparked by the DARPA Challenge in 2005 [1] and DARPA Urban Challenge in

H. Darweesh (✉)
Nagoya University, Furo-cho, Chikusa-ku, Nagoya, Japan
e-mail: hatem.darweesh@g.sp.m.is.nagoya-u.ac.jp

K. Takeda et al. (eds.), *Frontiers of Digital Transformation*,
https://doi.org/10.1007/978-981-15-1358-9_9

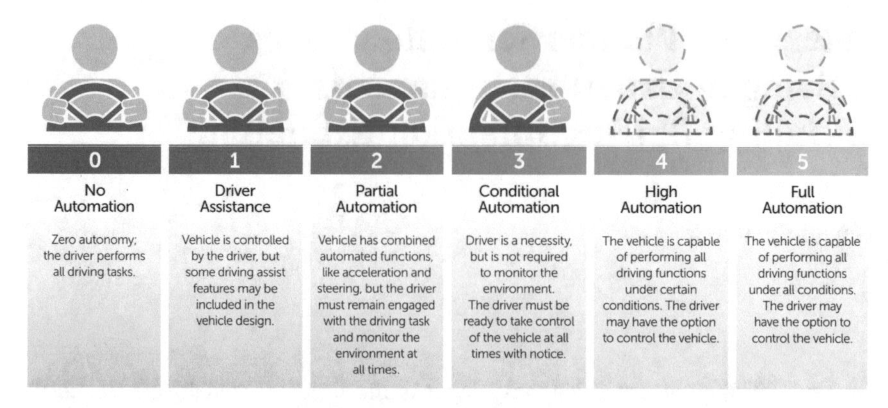

Fig. 1 Automation levels and milestones for autonomous driving, as defined by the Society of Automotive Engineers (SAE)

2007 [2]. However, after more than a decade of development, autonomous driving (AD) is still far from achieving its ultimate automation objectives. Figure 1 shows the milestones for autonomous driving, starting with the classical 100% human driver (no automation) to futuristic, 100% computer automated systems (full automation).[1]

Autonomous driving is still a difficult problem due to the vast number of possible situations that can occur in dynamic driving environments. Here, I propose a solution to one of the most challenging autonomous driving tasks, which is `planning`. My research objective is to develop an open-source framework for planning which can achieve the following goals:

- Fosters international collaboration
- Provides a complete autonomous driving software package
- Integrates the various planning modules so that they are usable in most autonomous systems
- Supports a broad range of platforms, such as stand alone APIs and within ROS
- Supports a broad range of maps, so that the system will support standard and open-source map formats
- Supports a broad range of environments, such as indoors, outdoors, structured streets, and off-road
- Incorporates intention awareness

[1]U.S. Department of Transportation, Automated Driving Systems 2.0, A Vision for Safety, https://www.nhtsa.gov/, [Online; accessed December 2019].

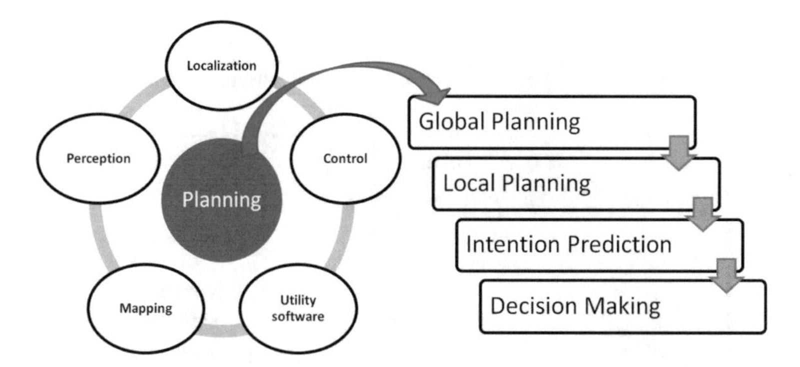

Fig. 2 Level 4+ autonomous driving software stack, showing the complexity of the planning part

1.1 Problem Definition

Planning consists of multiple modules as shown in Fig. 2. Integrating these together correctly is a challenging task.

Another challenge facing autonomous driving is the complex social interaction scenarios. In addition, location and culture difference pose additional challenges:

- In some countries, vehicles travel on the left side of the road, while in others on the right side
- Infrastructure differs from country to country and even from city to city
- Driving rules and habits differ
- Social interaction rules are different
- Traffic laws are different.

1.2 Proposed Solution

1.2.1 OpenPlanner: An Open-Source, Integrated Planner

The implementation of the open-source, integrated planner introduced in this work is called `OpenPlanner`. Its architecture is illustrated in Fig. 3. It includes a global planner that generates global reference paths using a vector (road network) map. The local planner then uses this global path to generate an obstacle-free, local trajectory from a set of sampled roll-outs. It uses various costs, such as collision, traffic rules, transition, and distance from center, to select the optimal trajectory. An intention and trajectory estimator calculates the probabilities associated with other vehicles' intentions and trajectories, while the behavior generator uses predefined traffic rules and sensor data to function as a decision-maker.

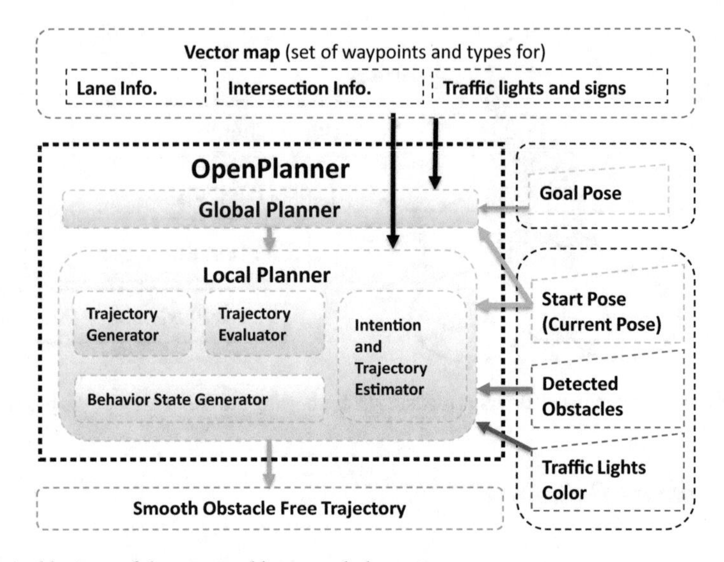

Fig. 3 Architecture of the proposed integrated planner

`OpenPlanner` has been used by many international research teams, which have made diverse and innovative contributions by applying it to a wide range of autonomous driving planning problems around the World, for example:

- **The Roboat project**: Using a fleet of autonomous boats [11] on Amsterdam's canals as transportation solution[2]
- **Autonomous driving map editor**: Road network map editor developed at the University of Konkuk, Korea [10]
- **ADAS Demo**: An Advanced Driver Assistance System (ADAS) demo for a major Hong Kong-based technology company

The integrated planner contributions could be seen in Table 1. It compares Open-Planner to the top open-source planners currently available.

1.2.2 Trajectory and Intention Estimation

I have developed a novel method of estimating the probabilities of the intentions and trajectories of surrounding vehicles, using an existing behavior planner with a particle filter. This estimation process is a very important function of the planner, allowing it to handle complex traffic situations.

[2]Roboat Project, MIT and AMS, https://roboat.org/, [Online; accessed December 2019].

Table 1 Comparison of OpenPlanner with leading open-source planners

	Planner					
Feature	Open Motion Planning Library (OMPL)	ROS navigation stack	Open robotics design & Control Open-RDC	Mobile robot programming toolkit MRPT	Apollo AD planner	Open planner
Platform support	Robots	Robots	Robots	Robots	Autonomous vehicles	Robots + Autonomous vehicles
Environment support	Indoor	Indoor, sidewalk	Indoor, sidewalk	Indoor	Pre-driven structured roads	Indoor, outdoor, public road, highway
Library APIs	Independent	–	–	Independent	–	Independent
ROS support	Yes	Yes	Yes	Via bridge	–	Yes
Map support	Cost map + point cloud	Cost map	Cost map	Cost map + point cloud	Customized OpenDRIVE	Vector Map, KML, OpenDRIVE, Lanelet2
Global planning	Yes	–	–	–	Yes	Yes
Local planning	Yes	Yes	Yes	Yes	Yes	Yes
Behavior planning	Custom	–	–	–	Yes	Yes

1.3 Chapter Structure

In Sect. 2, the components of OpenPlanner will be explained in detail, with intention and trajectory estimation discussed at length in Sect. 3. The relationship between OpenPlanner and the real-world data circulation is explained in Sect. 4. Finally, in Sect. 5, this chapter is concluded and future work is introduced.

2 Integrated Planner for Autonomous Navigation

The open-source, integrated planner introduced in this section can be used for autonomous navigation of mobile robots in general, including autonomous driving applications. It is designed to use road network map items such as lanes, traffic lights, traffic signs, intersections, and stop lines, which is one of its main advantages over other open-source as shown in Table 1. In this section, different components of the planner and experimental results are introduced.

2.1 Global Planning

The global planner handles path routing. It takes the vector map, a start position and a goal position as input and then finds the shortest or lowest cost path using dynamic programming [9]. The global planner used by OpenPlanner can support complicated vector maps. Dynamic programming is used to find the optimal path from start to goal positions as in Fig. 4.

2.2 Trajectory Planning

A local trajectory planner is a set of functionality that generates a smooth trajectory which can be tracked by path-following algorithms, such as Pure Pursuit [3]. For OpenPlanner, Roll-out generation approach in Fig. 5 is adapted, in which the behavior generator can demand a re-plan at any time to generate fresh, smooth, roll-out trajectories. The sampled roll-outs are divided into three sections as shown in Fig. 6.

2.3 Behavior Planning

The behavior state generation module of OpenPlanner functions as the decision-maker of the system. It is a finite state machine in which each state represents a traffic situation. Transitions between states are controlled by intermediate parameters calculated using current traffic information and pre-programmed traffic rules. Figure 7 shows the currently available states in the OpenPlanner system.

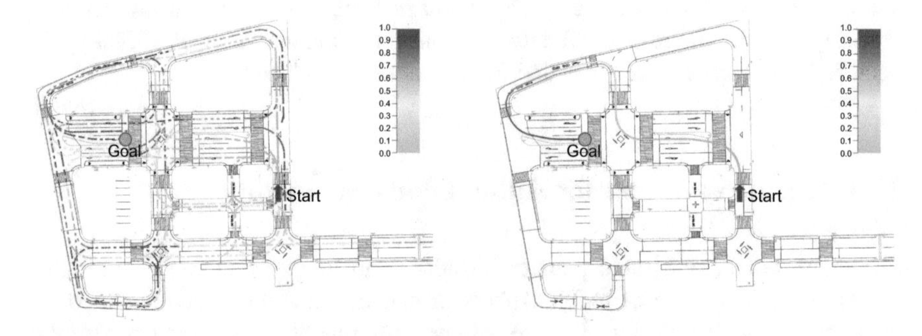

Fig. 4 Searching the map for the shortest path. Line color indicates route cost. Cost is represented by distance; green color is the closest distance to Start and red color is the max distance reached at the Goal position

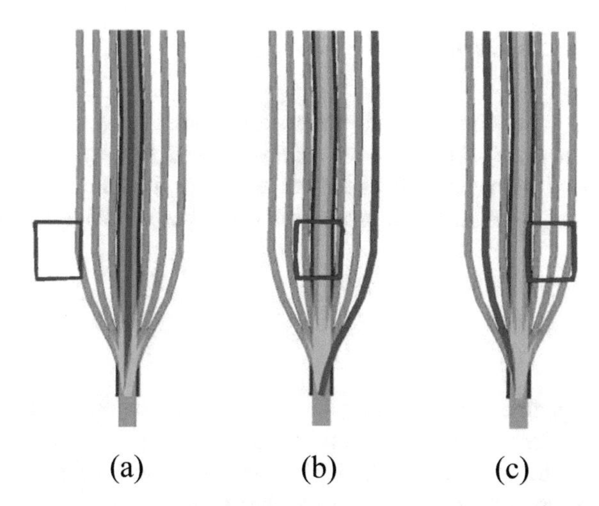

Fig. 5 Local Planner in action, in **a** the central trajectory is free, in **b** obstacle blocks the central trajectory so the most feasible one is the right-most trajectory, and in **c** the most feasible one is the second trajectory on the left

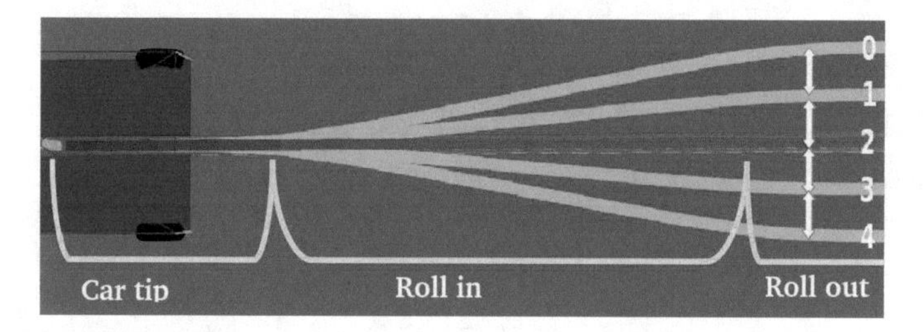

Fig. 6 Sections for generating roll outs

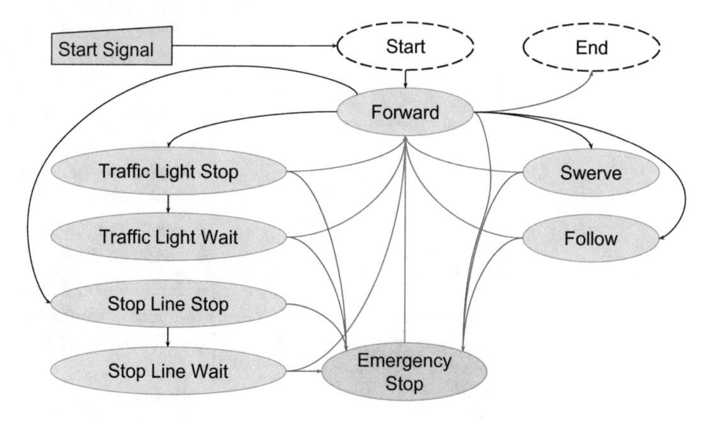

Fig. 7 Current system behavior states

2.4 Experiment Results

In this work, multiple robotics platforms are used, such as Ackerman-based steering robot, differential drive robot, and a real vehicle. These platforms are used in several experimental environments such as indoor, outdoor, structured public roads, and simulation environment. Table 2 shows the conducted experiments' maps and results.

2.5 Conclusion

The integrated planner was able to achieve the design objectives by successfully generating plans for multiple platforms in various dynamic environments. Also it targeted most of the open-source challenges we aimed for. Figure 8 shows the challenges and how OpenPlanner tackled these challenges.

As a result of providing this work as open-source, continuous feed back from the open-source community is received. There are diverse and innovative contributions

Table 2 Experimental results for proposed integrated planner

Experiment	Environment	Results
Simulation	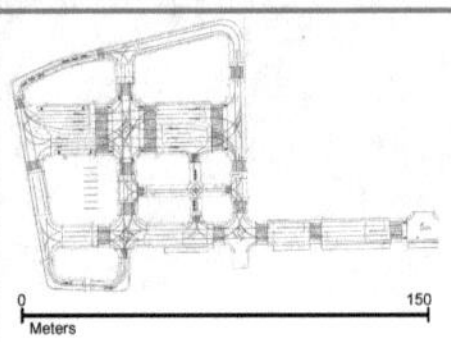	• The planner can correctly generate trajectories and select a suitable behavior • The global planner was able to find the shortest path on a complex map
Nagoya University	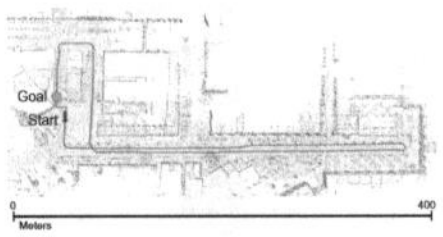	• The planner is able to perform in real time at a minimum 10 Hz, with an average of 100 objects detected • The planner is able to generate smooth trajectories and avoid obstacles even in a narrow passage • The planner is able to handle different traffic situations
Tsukuba Challenge	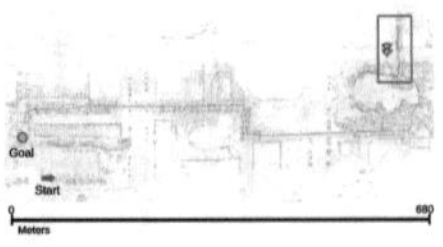	• The planner achieves automatic rerouting even when perception-based behavior is not available • The planner successfully navigates through a busy, unfamiliar environment

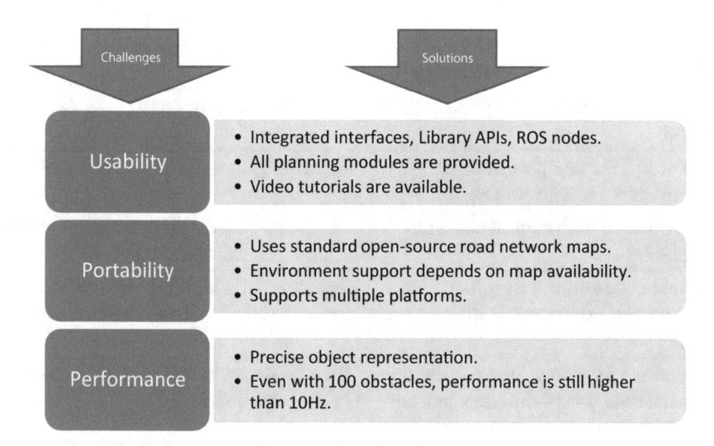

Fig. 8 Solution provided by OpenPlanner for the main challenges of developing an open-source planner for Autonomous Navigation systems

using OpenPlanner to a wide range of planning problems from around the World, for example:

- **The Roboat project**: MIT & AMS, Netherlands [11]
- **Autonomous driving open-source map editor**: Developed at the University of Konkuk, Korea [10]
- **Campus self driving carts**: University of California San Diego, USA [6]
- **ADAS Demo**: An Advanced Driver Assistance System (ADAS) demo, HKPC Hong Kong[3]

3 Behavior Planner Based Intention and Trajectory Estimation

Predicting with a high level of confidence what other vehicles are doing is essential for autonomous driving, and is one of the basic functions of a successful planner. Actions of other vehicles also include their probable intentions and future trajectories. The planner uses this information to generate suitable plans, i.e., ego-vehicle actions and trajectories.

The proposed solution is an intention and trajectory probability estimation algorithm, which utilizes a behavior planner [4] working in passive mode, wrapped in a multi-cue particle filter [7] for uncertainty modeling.

The main contribution is the development of a new method for estimating the intentions and trajectories of surrounding vehicles which can accurately handle most

[3] HKPC, HKPC web page, https://www.hkpc.org/en/, [Online; accessed December 2019].

Table 3 Issues with conventional particle filter estimation algorithms and contributions of the proposed estimation method by addressing these problems

Challenges using a conventional particle filter for estimation	Contributions of proposed approach
Difficult to model the complex motion parameters of surrounding vehicles	Behavior planner is used as the motion model
The state consists of both continuous and discrete variables	Multi-cue particle filter is used with variable confidence factors
Since the state dimensions are high, thousands of particles are needed to capture the posterior distribution	Separate particle filters are used for each intention and trajectory combination
Multi-cue particle filter uses joint distribution to aggregate the weights of different sensing cues, leading to particle deprivation	Weighted sum of different cues is used instead of the joint distribution

complex urban driving situations in real time. This is accomplished by using a behavior planner as the complex motion model, and integrating it with a non-parametric probabilistic filter (a multi-cue particle filter) to handle uncertainty. Table 3 highlights the main issues with the conventional approach and shows how these problems are solved.

3.1 Solution Approach

Figure 9 shows the system architecture of the proposed intention and trajectory estimation system. The proposed algorithm consists of three main parts; a trajectory extractor, a passive behavior planner, and multi-cue particle filters.

3.2 Passive Behavior Planner

The passive behavior planner is a stripped-out version of the integrated planner described in Sect. 2. The integrated planner is so flexible that unnecessary modules such as the Global Planner, Intention Estimation, and Object Tracking could be eliminated. It is called a passive planner because no feedback is available, thus the control signal does not have a direct impact on the observed state, and it has become similar to an open-loop planner.

Figure 10a shows the estimated intention states. For each intention state in Fig. 10a, there should exist a behavior that models the intention inside the passive behavior planner, Fig. 10b. Multiple behaviors can model single intention and vice versa.

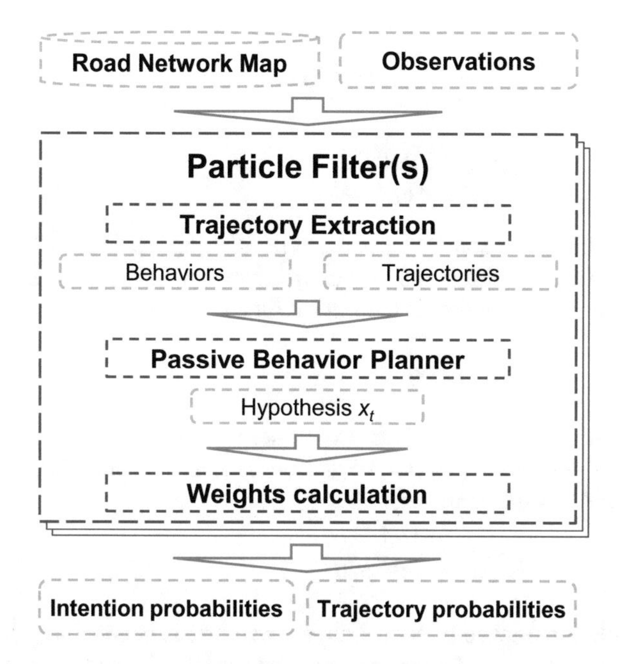

Fig. 9 Proposed intention and trajectory estimation system. Multi-cue particle filters use a passive behavior planner as a motion model

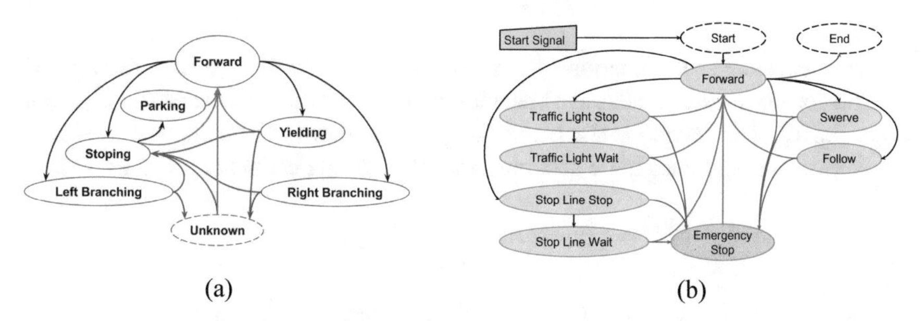

Fig. 10 Comparison of intentions to be estimated (**a**), to behaviors modeled by the passive behavior planner (**b**)

3.3 Uncertainty Modeling Using Particle Filter

The basic idea behind particle filtering is to approximate the belief state $b(x_t)$ at time t by a set of weighted samples χ_t as in Eq. (1) [5]. Here, x_t is the state vector, where z is the observation vector, χ_t^i means the ith sample of state x_t. Equation (1) enables computing the posterior probability using importance sampling. Because it is hard to sample from the target distribution, importance distribution $q(x)$ is used, then weighted according to Eq. (2).

$$P(x_t|z_{1:t}) \approx \sum_{i=1}^{N_s} w_t^i \delta(x_t, \chi_t^i) \tag{1}$$

$$w^i \propto \frac{P(x_{1:t}^i|z_{1:t})}{q(x_{1:t}^i|z_{1:t})} \tag{2}$$

In particle filter, the sample of a posterior distribution is called particles and denoted as χ in Eq. (3) with M the maximum number of particles. Here, each particle $x_t^{[m]}$ is a concrete instantiation of the state at time t.

$$\chi_t := x_t^{[1]}, x_t^{[1]}, \ldots, x_t^{[m]} \tag{3}$$

Including the hypothesis x_t in the particle set χ_t requires that it is proportional to the Bayes filter posterior belief $b(x_t) = P(x_t|z_t, u_t)$ as in relation (4), where z is the observation and u is the control signal.

$$x_t^{[m]} \propto P(x_t|Z_t, u_t) \tag{4}$$

The particle importance factor (particle's weights) is denoted as $w_t^{[m]}$, which is the probability of the measurement z_t under the particle $x_t^{[m]}$ and is given by Eq. (5).

$$w_t^{[m]} = P(z_t|x_t^{[m]}) \tag{5}$$

Here, two main modifications are introduced to the original particle filter in [8]. The first one is the use of a Passive behavior planner as the state transition distribution, as in Eq. (6). The second is the Multi-cue particle filter to allow the use of several weak measurement cues to be accumulated into a strong estimator, as in Eq. (7).

$$x_t^{[m]} \simeq x(z_t, x_{t-1}, M) \tag{6}$$

$$\begin{aligned} w_t^{[m]} = p(z_t|x_t^{[m]}) = \alpha_p \cdot P(z_{p,t}|x_t^{[m]}) + \alpha_d \cdot p(z_{d,t}|x_t^{[m]}) + \alpha_v \cdot p(z_{v,t}|x_t^{[m]}) \\ + \alpha_a \cdot p(z_{a,t}|x_t^{[m]}) + \alpha_s \cdot p(z_{s,t}|x_t^{[m]}) \end{aligned} \tag{7}$$

3.4 Estimation Algorithm

The proposed algorithm consists of three steps; trajectory extraction (initialization), particle sampling, and measurement update (weight calculation).

For the detected vehicle, all possible driving trajectories from the road network map is extracted. Then two additional trajectories to represent branching right and branching left are added.

The next step is the particle sampling to create hypothesis state x_t^m using the state transition $p(x_t|u_t, x_{t-1}^{[m]})$. State transition probability follows the motion assumed for each particle which is represented by the passive behavior planner working as an expert driver.

Finally, in the weight calculation step, the sampled particles are filtered by calculating how far the hypothesis distribution is from the measurement sensing cues. The weight for each particle is calculated against the sensing data, such as position, direction, velocity, acceleration, and turn signal information using Eq. (8).

$$w_t^{[m]} = \alpha_p \cdot w_{p,t}^{[m]} + \alpha_d \cdot w_{d,t}^{[m]} + \alpha_v \cdot w_{v,t}^{[m]} + \alpha_a \cdot w_{a,t}^{[m]} + \alpha_s \cdot w_{s,t}^{[m]} \tag{8}$$

3.5 Evaluation Results

The proposed intention and trajectory estimation system was evaluated using multiple simulated driving situations. The objective of these evaluations is to demonstrate that the proposed method can accurately estimate trajectories and intention probabilities in various driving scenarios, such as three-way intersection, four-way intersection, intersection with and without stop signs, and bus stops. These driving scenarios are depicted from the National Highway Traffic Safety Administration (NHTSA) report.[4]

The results in Table 4 show that the proposed method has the ability to accurately discriminate between various possible intentions and trajectories in a variety of complex driving situations.

4 Real-World Data Circulation and Social Impact

The idea of RWDC is based on openly creating, organizing, and sharing data, which capture the problem enabling teams to explore more possibility offline, analyzing existing solutions, and developing new ones based on the openly shared data. After development, another data set is created and shared which helps bench-marking the problem's solution. Data circulation creates a modern and unique way for international collaboration to tackle the most challenging technological problems. The work introduced in this chapter fits perfectly to the definition and support the RWDC concept. Section 4.1 shows how the integrated planner (OpenPlanner) relates to the RWDC concept. Additional project developed under the supervision of the RWDC Leading program at Nagoya University is introduced in Sect. 4.2 as another example of the utilization of the data circulation concept.

[4]U.S. Department of Transportation, Pre-crash Scenario Typology for Crash Avoidance Research, https://www.nhtsa.gov/, [Online; accessed December 2019].

Table 4 Experimental results for intention and trajectory estimation

Experiment	Environment	Results
Three way intersection	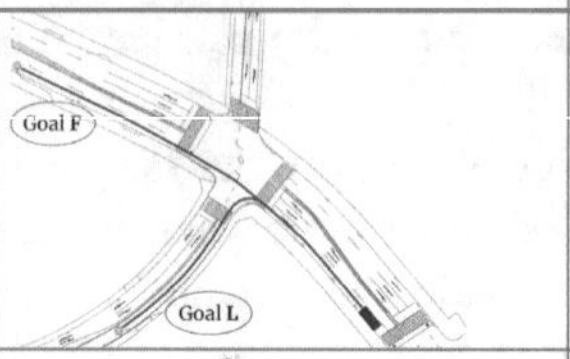	• The estimator successfully assigns higher probabilities to the correct intention and trajectory for both Goal **F** and Goal **L**
Four way intersection	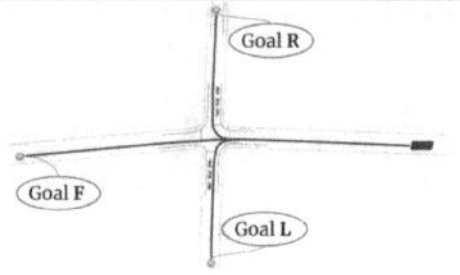	• The method shows successful estimation of intention and trajectory when there is sufficient sensing information • The method is tested in different situations, such as the vehicle stopping and not stopping at a stop line, and using or not using a turn signal • There was only one situation in which the estimator could not predict the trajectory, which was when the vehicle stopped at the stop line and no turn signal information was available. However, even the best human driver can't predict where the vehicle will go next in this situation
Bus stop (parking & yielding)	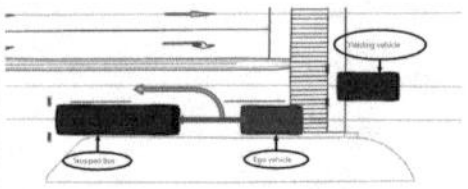	• The proposed method successfully estimates the parking intention of the bus and the intention of the vehicle in the other lane as to whether it would or would not yield. This is important because the planner needs to decide whether to wait behind the bus or to pass it

4.1 Autonomous Driving Planning

The introduced integrated planner utilizes the RWDC concept in two perspectives. First, data driven which is illustrated in Fig. 11. The second, as an open-source project, which is shown in Fig. 12.

The development of OpenPlanner contributed to the society as an open-source application. It has so far achieved the following:

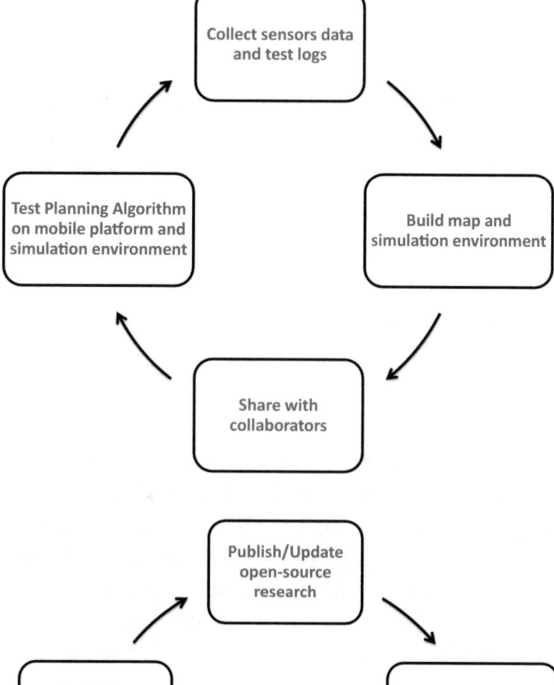

Fig. 11 Relationship between OpenPlanner and RWDC

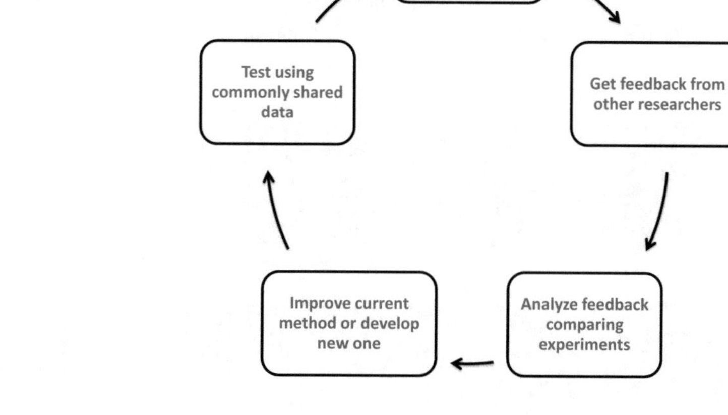

Fig. 12 OpenPlanner and other open-source code projects as examples of RWDC

- Hundreds of users, as well as feedback from the autonomous mobility community
- International collaboration with several teams in multiple countries, working on different goals
- Experiment and data sharing between development teams to improve the platform and create safer autonomous driving systems
- Use of the planner in multiple projects which directly improve daily life, such as the Roboat project.

4.2 Automated Road Network Mapping (ASSURE Maps)

Accurate road network maps are an essential part of reliable autonomous driving systems. ASSURE Maps project focused on developing a method of automatically building High Definition (HD) road network maps for autonomous vehicles. The goals of the ASSURE Maps project (and the origin of the project's name) were as follows:

- **A**ccurate road network maps
- **S**ecure cloud service
- **S**mart mapping tools
- **U**pdated maps
- **R**ich details
- **E**valuated results

The ASSURE Maps system consists of two main modules. The first module uses Automatic Map Generation **(AMG)** APIs, and the second uses Smart Mapping Tools **(SMT)**, as shown in Fig. 13. ASSURE's AMG APIs function as the internal engine that loads data logs (LiDAR, camera images, GPS data and odometry) and extracts map semantic information. The smart mapping tools module functions as a review tool which helps users control the output of the AMG APIs and create data sets for the system's machine learning-based components.

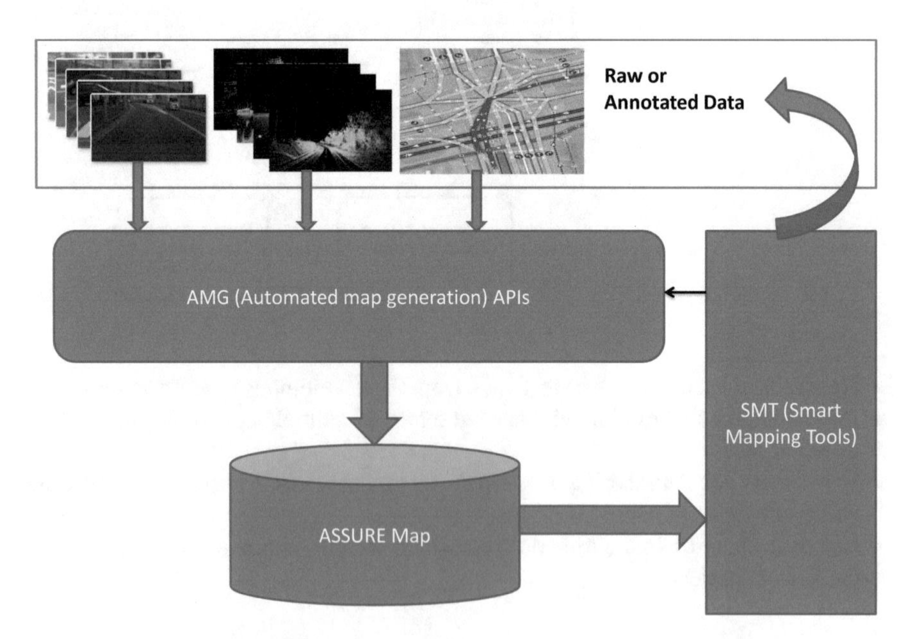

Fig. 13 ASSURE Maps Architecture

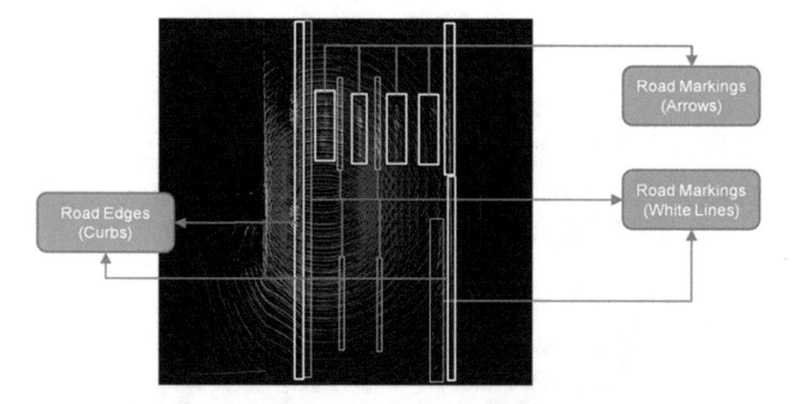

Fig. 14 LiDAR-based detection of the map data (curbs, markings, and lines)

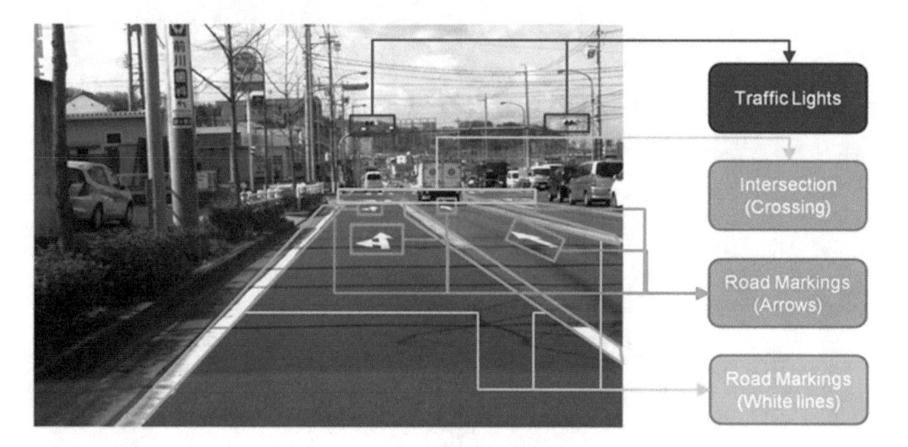

Fig. 15 Visual detection of map data (traffic lights, intersections, markings, and lines)

Experimental results in Fig. 14 show the extracted map items from the LiDAR data. The AMG APIs successfully detect curbs, lines, and markings. Figure 15 shows the detection of traffic light, intersections, markings, and lines using camera images only.

The relationship between ASSURE Maps and RWDC is twofold. The first is the data circulation that occurs at the core of the system development process. Figure 16 shows how data are utilized within the ASSURE mapping system. The second is based on the business model. Figure 17 shows the utilization of customer's data to improve the detection and mapping algorithms.

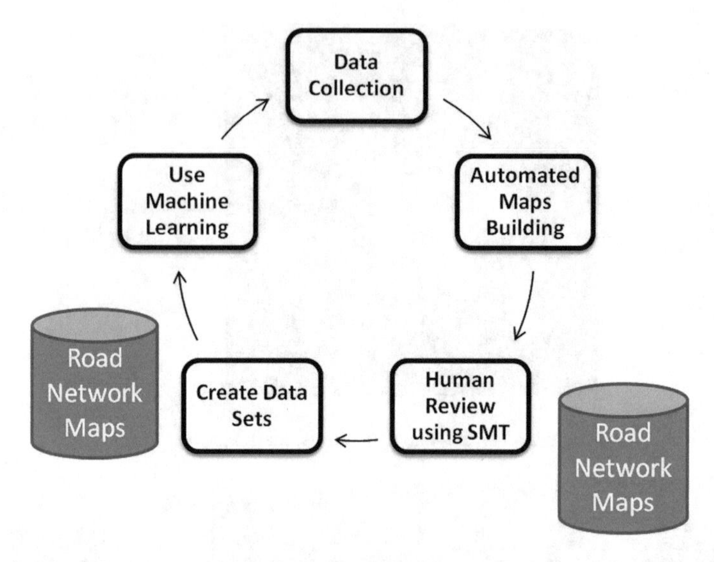

Fig. 16 Relationship between ASSURE maps and RWDC from a system development perspective

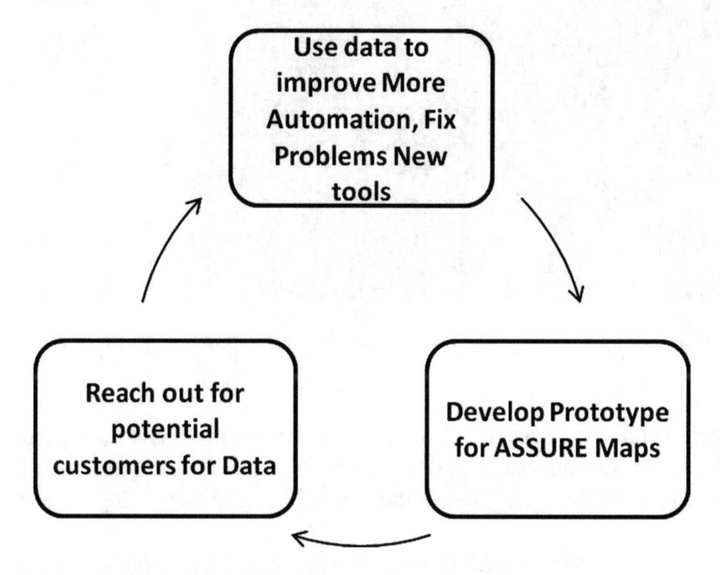

Fig. 17 Relationship between ASSURE Maps and RWDC from a business perspective

5 Conclusion and Future Work

In this chapter, the development of a complete, integrated, open-source planner for autonomous driving was introduced. The implementation of this planner, **OpenPlanner**, is open-source, so the robotics community can freely take advantage of it, use it, modify it, and build on it. OpenPlanner relies heavily on precisely created road network maps, which improve autonomous driving safety. Multiple experiments were conducted to show the planner functionalities. Continuous feedback from the open-source community indicates that OpenPlanner is useful not only for autonomous driving but also for a wide range of robotics applications.

In addition, intention and trajectory estimation method was introduced for predicting the actions of surrounding vehicles by associating a probability with each intention and trajectory. This is a very important step before decision-making in an autonomous driving system's planning process. A behavior planner is used to model the expected motion of surrounding vehicles, and particle filters are associated with each intention and trajectory. The results show that the proposed method has the ability to accurately discriminate between various possible intentions and trajectories in a variety of complex driving situations.

Achieving a speed of more than 30 km/h on a public road is one of our future goals. Another idea is to use MDP to calculate optimal motion actions rather than the current optimization method. Finally, for intention estimation filtering, using other probabilistic methods such as an HMM could improve performance dramatically.

Acknowledgments This work was supported by the OPERA project of the Japan Science and Technology Agency and by Nagoya University.

References

1. Buehler M, Iagnemma K, Singh S (2007) The 2005 DARPA grand challenge: the great robot race, 1st edn. Springer
2. Buehler M, Iagnemma K, Singh S (2009) The DARPA urban challenge: autonomous vehicles in city traffic, 1st edn. Springer
3. Coulter RC (1992) Implementation of the pure pursuit path tracking algorithm. Carnegie-Mellon University, Technical report
4. Darweesh H, Takeuchi E, Takeda K, Ninomiya Y, Sujiwo A, Morales LY, Akai N, Tomizawa T, Kato S (2017) Open source integrated planner for autonomous navigation in highly dynamic environments. J Robot Mechatron 29(4):668–684
5. Ghahramani Z (2001) An introduction to hidden Markov models and Bayesian networks. In: Hidden Markov models: applications in computer vision. World Scientific, pp 9–41
6. Paz D, Lai PJ, Harish S, Zhang H, Chan N, Hu C, Binnani S, Christensen H (2019) Lessons learned from deploying autonomous vehicles at UC San Diego. EasyChair Preprint no. 1295, EasyChair
7. Shen C, Van den Hengel A, Dick A (2003) Probabilistic multiple cue integration for particle filter based tracking. In: Proceedings of VIIth digital image computing, techniques and applications, pp 10–12
8. Thrun S, Burgard W, Fox D (2005) Probabilistic robotics. MIT Press

9. Thrun S, Montemerlo M, Dahlkamp H, Stavens D, Aron A, Diebel J, Fong P, Gale J, Halpenny M, Hoffmann G et al (2006) Stanley: the robot that won the DARPA grand challenge. J Field Robot 23(9):661–692
10. Tun WN, Kim S, Lee JW, Darweesh H (2019) Open-source tool of vector map for path planning in autoware autonomous driving software. In: Proceedings of 2019 IEEE international conference on big data and smart computing, pp 1–3
11. Wang W, Gheneti B, Mateos LA, Duarte F, Ratti C, Rus D (2019) Roboat: an autonomous surface vehicle for urban waterways. In: Proceedings of 2019 IEEE/RSJ international conference on intelligent robots and systems (IROS), pp 6340–6347

Direct Numerical Simulation on Turbulent/Non-turbulent Interface in Compressible Turbulent Boundary Layers

Xinxian Zhang

Abstract Direct numerical simulations for compressible temporally evolving turbulent boundary layers at Mach number of $M = 0.8$ and 1.6 are performed to investigate the turbulent/non-turbulent interface (TNTI) layer. The usual computational grid size determined based solely on the wall unit is insufficient near the TNTI, which results in spiky patterns in the outer edge of the TNTI layer and thicker TNTI layer thickness. With higher resolution direct numerical simulation (DNS), where the resolution is determined based on both the wall unit and the smallest length scale of turbulence underneath the TNTI layer, we investigate the characteristics of the TNTI layer in the compressible turbulent boundary layers. The thickness of the layer is found to be about 15 times of the Kolmogorov scale η_I in turbulence near the TNTI layer. The mass transport within the TNTI layer is well predicted by an entrainment model based on a single vortex originally developed for incompressible flows. The scalar dissipation rate near the TNTI is found to depend on the TNTI orientation: it is larger near the TNTI facing the downstream direction (leading edge). Finally, the real-world data circulation (RWDC) in the present study is explained. Besides, the contributions of this work to society are also discussed.

1 Background

1.1 Turbulent Boundary Layer

Turbulent boundary layer (TBL) plays an important role in many engineering applications and geophysical flows. For example, the development and separation of turbulent boundary layer can significantly affect the lift, drag, and also the instability of the aircraft and vehicle. Passive scalar is also important to be investigated for

X. Zhang (✉)
School of Aeronautic Science and Engineering, Beihang University, 37 Xueyuan Road, Haidian District, Beijing 100191, P. R. China
e-mail: zhangxinxian@buaa.edu.cn

K. Takeda et al. (eds.), *Frontiers of Digital Transformation*,
https://doi.org/10.1007/978-981-15-1358-9_10

studying turbulent boundary layer, which is a diffusive contaminant in a fluid flow. Understanding the behavior of passive scalar is a necessary step in understanding turbulent mixing, chemical reaction, or combustion [41]. For example, passive scalar can show the mixing degree of two different flows in a reaction flow.

1.2 Turbulent/Non-turbulent Interface

Over the past few decades, a large number of studies [15, 27, 38] have been devoted to understanding TBL from various points of view. Turbulent boundary layers are known as highly intermittent flows, where both turbulent and non-turbulent (laminar) fluids coexist. The studies show that the turbulent and non-turbulent flows are separated by an apparent boundary. In 1928, Prandtl [26] firstly pointed out the existence of this sharp interface between turbulent and non-turbulent flows in the intermittent region, which is called turbulent/non-turbulent interface (TNTI). After decades, the existence of TNTI was firstly examined in a free shear layer by Corrsin and Kistler [6], and recent studies [1, 29] have revealed that the TNTI is a thin layer with finite thickness. The turbulent and non-turbulent flow regions are separated by this TNTI layer, where flow properties, such as enstrophy, kinetic energy dissipation, and scalar concentration, sharply change in this layer so that they are adjusted between the turbulent and non-turbulent flows [28]. This layer is also important for the exchanges of substance, energy, and heat between turbulent and non-turbulent flow and is also related to the spatial development of turbulence [10]. Therefore, it is very important to understand the characteristics of TNTI.

The spatial distribution of turbulent fluids also plays an important role in scalar mixing in TBLs because turbulence can create small-scale fluctuating scalar fields, which enhances turbulent mixing at the molecular level. Modeling of turbulent mixing is crucial in numerical simulations of reacting flows [8] and combustions [35]. One of the key quantiles in the modeling of turbulent reacting flows is scalar dissipation rate, which strongly depends on the characteristics of turbulence [34]. Many models developed for simulating turbulent reacting flows contain the scalar dissipation rate as an unknown variable [5, 7, 17, 24]. The TNTI often appears near the interface that separates two streams with different chemical substances, thus, the mixing process near the TNTI can be important in chemically reacting flows [9, 42, 44], where the chemical reaction rate is strongly affected by turbulence. Therefore, it is also important to investigate the characteristics of scalar mixing near the TNTI.

2 Current Researches and Our Objectives

2.1 Current Researches

The TNTI appears in many canonical flows such as jets, wakes, and boundary layers. Recently, with the improvement of supercomputer resources and laser-based measurement techniques, many numerical simulations and experiments have been conducted to investigate the TNTI in canonical turbulent flows [29]. The flow properties near the TNTI in these flows have been investigated with the conditional statistics computed as a function of the distance from the TNTI [1]. The TNTI layer is found to consists of two (sub) layers with different dynamical characteristics as shown in Fig. 1. The outer part is called viscous superlayer (VSL), where viscous effects dominate vorticity evolution, while the region between the VSL and turbulent core region is called turbulent sublayer (TSL) [36], where the inviscid effects, such as vortex stretching, become important.

The phenomenon of mass transferred from the non-turbulent region to the turbulent region is called the turbulent entrainment process, by which turbulence spatially grows. As introduced in previous study [22], the entrainment caused by large-scale eddies is called engulfment, and the entrainment process caused by small-scale eddies near the TNTI is referred to as nibbling. The nibbling-type entrainment is aroused by the viscous diffusion of vorticity near the TNTI layer while the engulfment is described as the non-turbulent flow which is drawn into the turbulent side by large-scale eddies before acquiring vorticity. The dominant mechanism for the entrainment process has been argued for many years. Recent studies have suggested that the nibbling process is responsible for the entrainment mechanism, and large-scale features of turbulence impose the total entrainment rate [10, 22, 47].

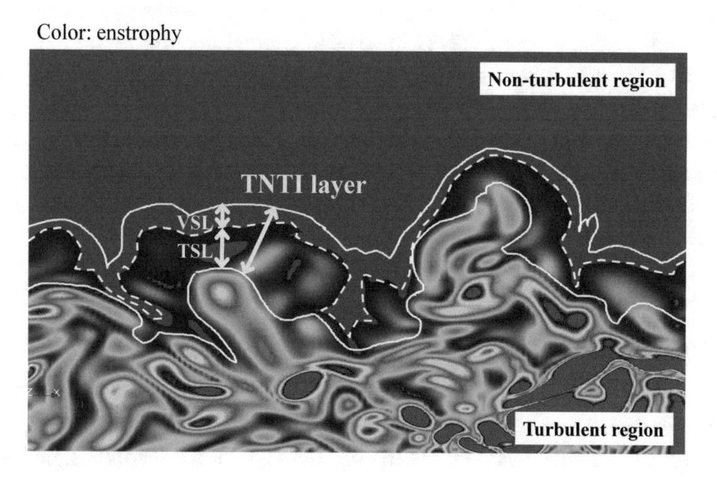

Fig. 1 The concept of inner structures in TNTI layer

The geometry of TNTI is also an important issue for understanding the entrainment process: large pockets structures on the TNTI interface can indraft the non-turbulent fluids into the turbulent region before acquiring vorticity (engulfment) if the TNTI interface is intensely folded [37]. It is doubtless that the complex geometry of the interface is highly related to the total entrainment rate because the total entrainment rate can be expressed as the surface integral of the local entrainment velocity (nibbling). Therefore, it may need more information to know the relation between nibbling and engulfment as mentioned by Borrell and Jiménez [2]. Some recent studies have revealed the influence of large-scale structures on the geometry of TNTI in the boundary layer [20, 32], which can make differences in the entrainment process between the TBLs and free shear flows because the large-scale motions depend on flow types.

Furthermore, the turbulent flow under the TNTI layer contains eddies with a wide range of scales, and all length scales can affect the properties and geometry of the TNTI layer. Therefore, motions from the smallest to the largest scales need to be captured in measurement or simulations. Especially in direct numerical simulations (DNS), all scales should be resolved, and insufficient resolution can directly affect computational results. With the DNS, researchers are able to access three-dimensional data of all quantities, which is difficult to obtain in experiments especially in high-speed flows. However, the resolution of TNTI in DNS for TBLs has not been investigated.

Recently, the TNTI in TBLs have been studied in experiments [4, 25, 31, 32, 48] and direct numerical simulations [2, 11, 20, 32]. Some characteristics of the TNTI in TBLs are found to be similar to the ones in the free shear flows [30], e.g., vorticity jump in the TNTI layer, and fractal features of the TNTI. There are still few studies on the TNTI in the turbulent boundary layer compared with in free shear flows. Especially, TNTI studies in TBL have been done in incompressible flows.

In many aerospace engineering applications, the TBLs often develop in a transonic or supersonic free-stream, where compressibility is no longer negligible [33]. Compressibility effects on the TNTI have been studied in compressible mixing layers [12, 13, 23, 40]. There are only a few experimental studies on the TNTI in high-speed TBLs [48, 49], in which they conducted fractal analysis on the TNTI of supersonic turbulent boundary layer with the experimental data. However, the flow measurement near the TNTI is very limited and difficult especially in high-speed flows so that many characteristics about TNTI in compressible TBLs have not been investigated.

2.2 *Objectives*

Even though there are already some DNS studies on the TNTI in TBL [2, 11, 20], grid setting has not been evaluated with consideration of the resolution on the TNTI in TBL. This is because most DNS studies for TBL focus on the near-wall region, where the grid spacing is carefully considered in usual DNS for TBL, while the intermittent region has not been fully considered.

The TNTI has been extensively studied in recent studies on free shear flows and some similar characteristics of the TNTI are also found in the TBL [30]. However, there are few studies on the TNTI in the TBL than in free shear flows, and many issues remain unclear for the entrainment process near TNTI layer. Although a high-speed regime is of great importance in realistic aerospace engineering applications, most studies on the TNTI have been done in incompressible flows. However, the TNTI in compressible turbulence is still less understood compared with the one in incompressible flows.

Understanding the characteristics of the TNTI is greatly important in modeling and predicting the spatial development of turbulence as well as the flow control based on the turbulent structures near the TNTI. As described above, it is important to investigate the TNTI in compressible TBLs. In this study, direct numerical simulations with two types of grid setting are performed for both the subsonic and supersonic turbulent boundary layers in order to investigate the spatial resolution effects on TNTI, compressibility effects, entrainment process, as well as the development of the high-speed turbulent boundary layers.

The main objectives in the present study are to

a. Develop the DNS code for the compressible TBLs with two types of grid setting.
b. Evaluate a reasonable grid setting for the TNTI study in compressible TBLs.
c. Investigate the compressibility effects on TNTI in compressible TBLs.
d. Elucidate the physical mechanism of the entrainment in compressible TBLs.

3 Main Achievements

3.1 High Resolution Simulation

DNS of subsonic and supersonic temporally evolving turbulent boundary layers are performed for studying the TNTI. Two different setups of the DNS are considered, where in one case the grid spacing is determined solely based on the wall unit (case C001 and C002) while the other case uses the computational grid small enough to resolve both the turbulent structures underneath the TNTI and near the wall (case F001 and F002). The global statistics are compared between the present DNS and previous studies, showing that the DNSs with both grids reproduce well the first- and second-order statistics of the fully-developed turbulent boundary layers. The visualization of temporally evolving turbulent boundary layer with high resolution for $M = 0.8$ (case F001) is shown in Fig. 2.

However, the spatial distribution of vorticity in the outer region is found to be very sensitive to the spatial resolution near the TNTI. At the present Reynolds number ($Re_\theta \approx 2200$), the DNS based on the grid size determined by the wall unit does not have sufficient resolutions near the TNTI. The lack of resolution results in spiky

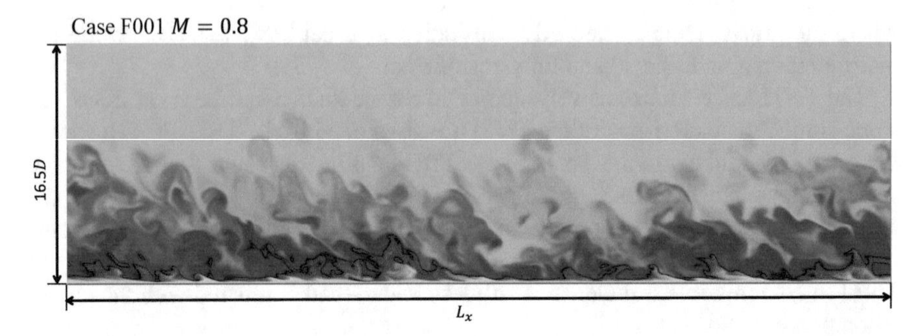

Fig. 2 Visualization of temporally evolving turbulent boundary layer for $M = 0.8$ case F001. Color represents passive scalar ϕ

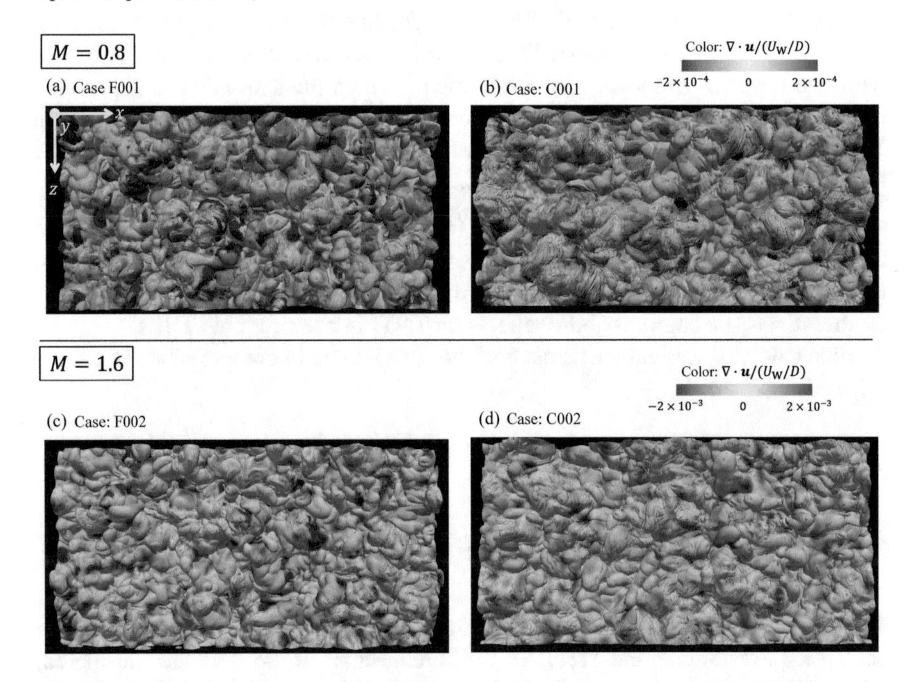

Fig. 3 Visualization of irrotational boundary forming at the outer edge of the TNTI layer for $M = 0.8$ (case F001, case C001) and $M = 1.6$ (case F002, case C002). Color represents dilatation $\nabla \cdot u$

patterns of the enstrophy isosurface used for detecting the outer edge of the TNTI layer (Fig. 3) and thicker TNTI layer thickness. This problem can be solved by increasing the number of the grid points, where a smoother enstrophy isosurface is similar to the previous studies of incompressible free shear flows obtained in the DNS with the grid small enough to resolve Kolmogorov scale in the turbulent core region below the TNTI.

3.2 TNTI in Compressible Turbulent Boundary Layers

Based on the 3D high-resolution DNS, the structure of the TNTI layer in compressible turbulent boundary layers can be investigated. The outer edge of TNTI layer, namely, irrotational boundary, is detected as an isosurface of vorticity as shown in Fig. 3a and c. The present results show that the thickness of the TNTI layer, defined with a large gradient of conditional mean vorticity magnitude, is about 15 times of the Kolmogorov scale η_{TI} in turbulence near the TNTI layer. The inner (sub)layers of the TNTI layer are detected based on the vorticity dynamics, where the TSL and VSL are found to have a thickness of 11–12η_{TI} and 4η_{TI}, respectively.

Even though the compressibility effects increase with Mach number, the conditional statistics confirm that the direct influences of compressibility are small near the TNTI layer, and the profiles of the conditional statistics are qualitatively similar between incompressible and compressible turbulent boundary layers. These structures of the TNTI layer and their thicknesses divided by the Kolmogorov scale are very similar to those found in incompressible free shear flows. The compressibility effects at the Mach numbers $M = 0.8$ and 1.6 are very small within the TNTI layer, which appears in the outer intermittent region.

The local entrainment process is studied with the propagation velocity of the enstrophy isosurface, which represents the speed at which non-turbulent fluids cross the outer edge of the TNTI layer. It has been shown that the compressibility effects are almost negligible for the propagation velocity, which is dominated by the viscous effects rather than a dilatational effect or baroclinic torque. The mean downward velocity is found in the non-turbulent region in the intermittent region, which is consistent with spatially evolving boundary layers [3, 20]. The mass entrainment rate per unit horizontal area of the temporal TBLs is consistent with the theoretical prediction [33] for the spatial compressible TBLs. This confirms that the dominant mechanism for the momentum transport, which is related to the TBL thickness growth, is not different between spatial and temporal compressible TBL as also found in incompressible TBLs [18]. Furthermore, the mass entrainment rate normalized by $u_\tau \rho_0$ at $M = 0.8$ also agrees well with experiments of spatially developing incompressible TBLs at various Reynolds numbers. Furthermore, the entrainment process across the TNTI layer is studied with the mass transport equation in the local coordinate system (x^{I}, t') which is moving with the outer edge of the TNTI layer. The statistics of the mass flux show that the mass within the VSL is transferred toward the TSL in the direction normal to the TNTI while the TSL is dominated by a tangential transfer. These mass fluxes within the VSL and TSL are compared with the single vortex model for the entrainment within the TNTI layer, which was proposed for incompressible flows [45] because of very small effects of the compressibility in the outer region of the turbulent boundary layer, the entrainment model given by a single vortex predicts the mass flux within the TNTI layer fairly well, which strongly suggests the connection between the entrainment process within the TNTI layer and the small-scale vortical structures found underneath the TNTI layer of the turbulent boundary layers.

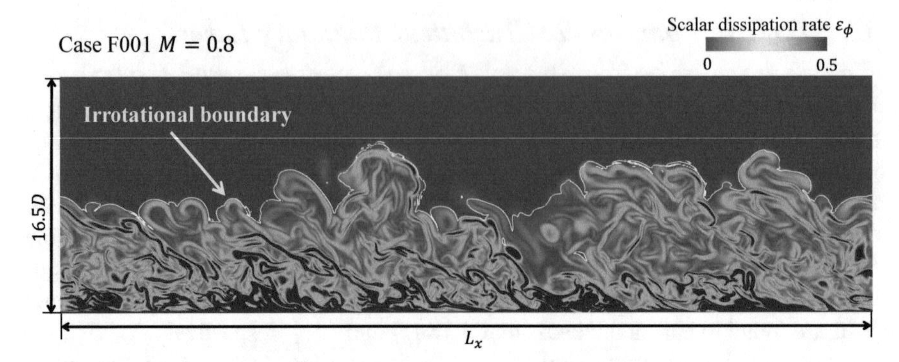

Fig. 4 Instantaneous profile of scalar dissipation rate (color contour) and irrotational boundary (white line) on x-y plane for case F001 ($M = 0.8$)

The irrotational boundary detected by an isosurface of passive scalar is also shown in this study, and the detected irrotational boundary shows an excellent agreement with the one detected by vorticity in visualization. It indicates passive scalar is also a good marker of turbulent fluids, which is easy to measure in experiments compared with vorticity. Conditional mean passive scalar also exhibits a sharp jump within the TNTI layer, and the highest conditional mean scalar dissipation rate appears near the boundary between the VSL and TSL. This indicates that the fluid locally entrained from non-turbulent side encounters the fluid coming from the turbulent side, where the difference in the passive scalar between these fluids creates large scalar gradients. It is also shown that the production rate of scalar gradient and enstrophy within the TNTI layer is as high as in the turbulent core region, and peaks in conditional averages of these quantities appear within the TNTI layer. Both visualization and conditional statistics show the dependence on the TNTI orientation for the scalar dissipation rate and the production rate of scalar gradient as shown in Fig. 4, both of which have a large value near the leading edge facing the downstream direction than the trailing edge facing the upstream direction. The production rate of scalar gradient within the TNTI layer of the trailing edge is comparable to the non-turbulent value, which causes a lower scalar dissipation rate near the trailing edge. These tendencies are explained from the difference in streamwise velocity between turbulent and non-turbulent fluids in a similar way to the TNTI orientation dependence of enstrophy production (rate) given for incompressible planar jets [43, 46].

4 Real-World Data Circulation

Real-World Data Circulation (RWDC) is a recent advanced new research field, which is often conducted for the real-world products or services in society. The information in RWDC is represented and analyzed in the form of data. By analyzing these data, people can create new designs or improve the old ones.

RWDC exists in virtually all fields related to our lives, including business, medical treatment, economics, education, and industry, and is highly related to the development and globalization of the world [16, 39]. In the real world, even though people get data or information from the fields they are interested in, it is difficult to know how to use these data to contribute to our life or improve our knowledge. That is the reason why we need to study RWDC.

Industries are of great importance in our daily life, from the plastic bag for food to the aircraft by which people can travel everywhere on this earth. However, manufacturers cannot unilaterally create valuable products. In other words, techniques and user requirements are necessary for creating valuable products or services. The recent proposed fourth industrial revolution (Industry 4.0) [19], which has been developing with Information Technology (IT), has attracted a lot of attention. This industrial revolution tries to connect many fields together, e.g., the Internet of Things (IoT), Cyber-Physical System (CPS), information and communications technology (ICT), and Enterprise Architecture (EA) [21]. RWDC is one of the key points in the realization of Industry 4.0.

4.1 Relation Between RWDC and the Present Study

Due to technological limitations, it is still difficult for industry to create desired products or services based on current techniques. For advancing the techniques, it is necessary to deeply understand the fundamental aspects, namely, practical or industrial related science, which can help engineers essentially achieve a new stage of the technique.

Fluid dynamics plays a significant role in the performance of many industrial applications, such as aircraft and automobile. How to use the practical or industrial related science in fluid dynamics to improve industries can be studied by RWDC as shown in Fig. 5, which is a general concept of RWDC in fluid dynamics for industrial products design. Some products in industries may have some problems or users may not be satisfied with the products, but engineers may not be able to solve the problems and need more information from the theory or fundamental characteristics. Then, researchers conduct some simulations or experiments to acquire data for fundamental studies. Thirdly, these fundamental studies can be done with the acquired data, and it can contribute to the development of theories or empirical laws. Finally, theories or empirical laws can be directly used for industrial design. By these steps, RWDC builds bridges for scientific study and real-world product design.

Computational fluid dynamics (CFD) is a widely used method for studying fluid dynamics, which has been using to design many parts of the aircraft. The usage of CFD is growing rapidly due to the increases of computational resource [14], which also decreases the time and financial cost of aircraft design, because aerodynamics studies of aircraft highly depended on the wind tunnel experiments in the past, that takes a lot of time, human resources, and financial cost. Commercial CFD software (CFX, Fluent, Fastran and Comsol, and so on) are frequently used for flow simulations

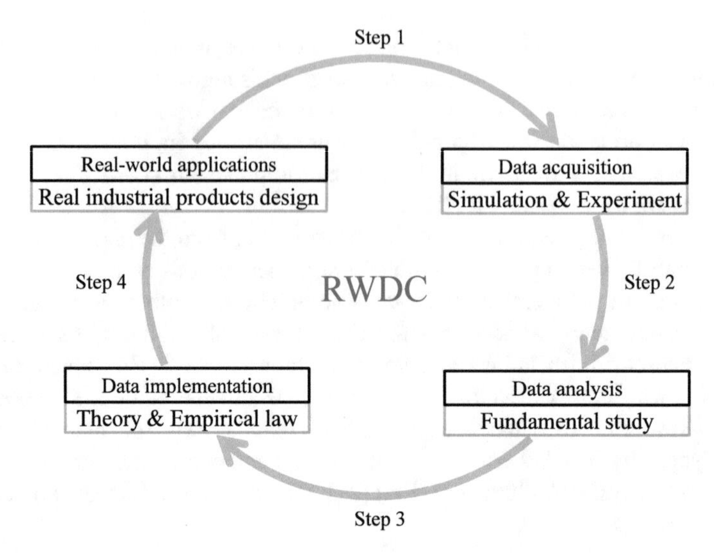

Fig. 5 The general concept of RWDC in fluid dynamics

in industrial applications, including many complex flows, e.g., combustion flow, high-speed flow, and flow with a heating wall. However, the turbulent flow simulations for industrial applications can only be done with the turbulent model due to the limited computational resource in the current stage. Even though these turbulent models excellently reduce the computational cost, it also brings some accuracy problems, especially in complex flows. Until now, a turbulent model is still one of the biggest problems in CFD community.

As mentioned in da Silva et al. [29], a better understanding of the TNTI layer is needed for predicting the chemical reaction flows or combustion flows, in which the properties essentially depends on the position and inner structures of TNTI layer. New computational algorithms and turbulent models near the TNTI layer should also be further considered because this layer separates the turbulent region from the non-turbulent region where the grid mesh used in applications near the TNTI is generally larger than the thickness of this layer.

The present study focuses on the TNTI layer in compressible turbulent boundary layers, which can be found in many engineering applications as described in Sect. 1 of this chapter, especially in aerospace engineering. The relation between RWDC and the present study can be explained by four steps as shown in Fig. 6. Firstly, the development of compressible TBLs can significantly affect the efficiency and instability of the aircraft and vehicles, and the many fundamental mechanisms about the development of compressible TBLs are still unclear as described in Sect. 1. These issues need to be investigated more. Then, direct numerical simulations of compressible TBLs are conducted for studying the fundamental characteristics of the TNTI layer and the entrainment process in compressible TBLs, which are strongly related to the development of compressible TBLs. Direct numerical simulations are conducted in

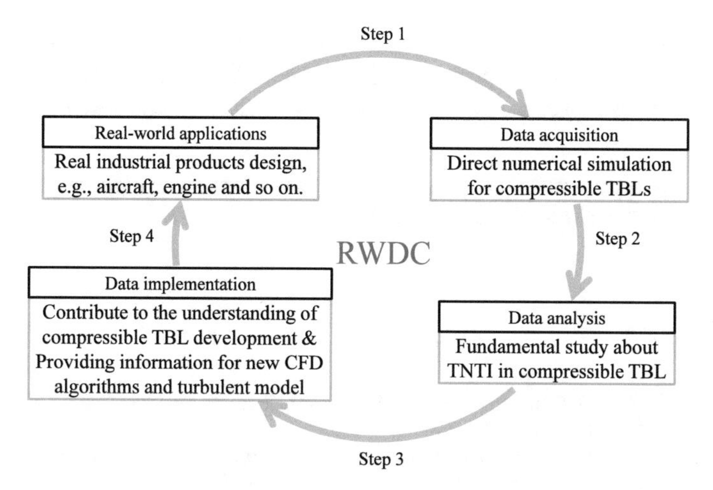

Fig. 6 Connection of RWDC and TNTI study in compressible TBLs

the present study because it directly solves all the turbulence motions with different scales without any turbulent model. Thirdly, the results about the TNTI layer in the present study can contribute to the understanding of the development of compressible TBLs, which provides information for the CFD algorithms and turbulent model development near the TNTI. Finally, these improvements in understanding of the development of compressible TBLs can be directly used in the design of real-world industrial products such as aircrafts and turbines.

4.2 *Contributions to the Society*

The present work can contribute to the understanding of the development of compressible TBLs, which can significantly contribute to design of industrial products from three points:

a. It can be used to improve the turbulent model and new CFD algorithms near the TNTI for compressible TBLs as described in Sect. 4.1. This will be a very big step if a suitable turbulent model or new CFD algorithms can be developed for industries because an improvement of CFD is an improvement for all the products designs which are related to the high-speed boundary layer. For example, the improvement of CFD in compressible TBLs can exactly improve the prediction of the flow around a high-speed vehicle, and it can be used to design a better vehicle with less drag and higher stability, subsequently better efficiency.
b. Besides, understanding the development of compressible TBLs is also helpful for developing related theories or empirical laws, that can be directly used in the industrial product design.

c. Furthermore, it also gives some ideas for flow control, e.g., it is possible to control the development of compressible TBL by controlling the large structures in the boundary layer because the large structures can affect properties of the TNTI in TBLs.

The work in the present study mainly focuses on the first and second steps of RWDC shown in Fig. 6, and contributes to the third step.

References

1. Bisset DK, Hunt JCR, Rogers MM (2002) The turbulent/non-turbulent interface bounding a far wake. J Fluid Mech 451:383–410
2. Borrell G, Jiménez J (2016) Properties of the turbulent/non-turbulent interface in boundary layers. J Fluid Mech 801:554–596
3. Chauhan K, Philip J, Marusic I (2014a) Scaling of the turbulent/non-turbulent interface in boundary layers. J Fluid Mech 751:298–328
4. Chauhan K, Philip J, de Silva CM, Hutchins N, Marusic I (2014b) The turbulent/non-turbulent interface and entrainment in a boundary layer. J Fluid Mech 742:119–151
5. Cleary MJ, Klimenko AY (2009) A generalised multiple mapping conditioning approach for turbulent combustion. Flow Turbul Combust 82(4):477–491
6. Corrsin S, Kistler AL (1955) Free-stream boundaries of turbulent flows. NACA Tech Rep (1244)
7. Curl RL (1963) Dispersed phase mixing: I. Theory and effects in simple reactors. AIChE J 9(2):175–181
8. Fox RO (2003) Computational models for turbulent reacting flows. Cambridge Univ, Pr
9. Gampert M, Kleinheinz K, Peters N, Pitsch H (2014) Experimental and numerical study of the scalar turbulent/non-turbulent interface layer in a jet flow. Flow Turbul Combust 92(1–2):429–449
10. Holzner M, Lüthi B (2011) Laminar superlayer at the turbulence boundary. Phys Rev Lett 106(13):134503
11. Ishihara T, Ogasawara H, Hunt JCR (2015) Analysis of conditional statistics obtained near the turbulent/non-turbulent interface of turbulent boundary layers. J Fluids Struct 53:50–57
12. Jahanbakhshi R, Madnia CK (2016) Entrainment in a compressible turbulent shear layer. J Fluid Mech 797:564–603
13. Jahanbakhshi R, Vaghefi NS, Madnia CK (2015) Baroclinic vorticity generation near the turbulent/non-turbulent interface in a compressible shear layer. Phys Fluids 27(10):105105
14. Jameson A, Ou K (2011) 50 years of transonic aircraft design. Proc Aero Sci 47(5):308–318
15. Jiménez J (2018) Coherent structures in wall-bounded turbulence. J Fluid Mech 842
16. Joung B (2017) Real-world data and recommended dosage of non-vitamin k oral anticoagulants for korean patients. Korean Circ J 47(6):833–841
17. Klimenko AY, Bilger RW (1999) Conditional moment closure for turbulent combustion. Prog Energy Combust Sci 25(6):595–687
18. Kozul M, Chung D (2014) Direct numerical simulation of the incompressible temporally developing turbulent boundary layer. In: Proceedings 19th Austral. Fluid Mechnical Conference, Austral. Fluid Mech. Soci
19. Lasi H, Fettke P, Kemper HG, Feld T, Hoffmann M (2014) Industry 4.0. Bus Inf Syst Eng 6(4):239–242
20. Lee J, Sung HJ, Zaki TA (2017) Signature of large-scale motions on turbulent/non-turbulent interface in boundary layers. J Fluid Mech 819:165–187
21. Lu Y (2017) Industry 4.0: a survey on technologies, applications and open research issues. J Ind Inf Integr 6:1–10

22. Mathew J, Basu AJ (2002) Some characteristics of entrainment at a cylindrical turbulence boundary. Phys Fluids 14(7):2065–2072
23. Mathew J, Ghosh S, Friedrich R (2016) Changes to invariants of the velocity gradient tensor at the turbulent-nonturbulent interface of compressible mixing layers. Int J Heat Fluid Flow 59:125–130
24. Meyer DW (2010) A new particle interaction mixing model for turbulent dispersion and turbulent reactive flows. Phys Fluids 22(3):035103
25. Philip J, Meneveau C, de Silva CM, Marusic I (2014) Multiscale analysis of fluxes at the turbulent/non-turbulent interface in high Reynolds number boundary layers. Phys Fluids 26(1):015105
26. Prandtl L (1928) Motion of fluids with very little viscosity. NACA Tech Memo 452
27. Robinson SK (1991) Coherent motions in the turbulent boundary layer. Annu Rev Fluid Mech 23(1):601–639
28. da Silva CB, Dos Reis RJN, Pereira JCF (2011) The intense vorticity structures near the turbulent/non-turbulent interface in a jet. J Fluid Mech 685:165–190
29. da Silva CB, Hunt JCR, Eames I, Westerweel J (2014a3) Interfacial layers between regions of different turbulence intensity. Annu Rev Fluid Mech 46:567–590
30. da Silva CB, Taveira RR, Borrell G (2014b) Characteristics of the turbulent/nonturbulent interface in boundary layers, jets and shear-free turbulence. J Phys Conf Ser 506:012015
31. de Silva CM, Philip J, Chauhan K, Meneveau C, Marusic I (2013) Multiscale geometry and scaling of the turbulent-nonturbulent interface in high Reynolds number boundary layers. Phys Rev Lett 111(4):044501
32. de Silva CM, Philip J, Hutchins N, Marusic I (2017) Interfaces of uniform momentum zones in turbulent boundary layers. J Fluid Mech 820:451–478
33. Smits AJ, Dussauge JP (2006) Turbulent shear layers in supersonic flow. AIP Pr
34. Su LK, Clemens NT (1999) Planar measurements of the full three-dimensional scalar dissipation rate in gas-phase turbulent flows. Exp Fluids 27(6):507–521
35. Tanaka S, Watanabe T, Nagata K (2019) Multi-particle model of coarse-grained scalar dissipation rate with volumetric tensor in turbulence. J Comput Phys 389:128–146
36. Taveira RR, da Silva CB (2014) Characteristics of the viscous superlayer in shear free turbulence and in planar turbulent jets. Phys Fluids 26(2):021702
37. Townsend AA (1976) The structure of turbulent shear flow. Cambridge Univ, Pr
38. Tsuji Y, Fransson JHM, Alfredsson PH, Johansson AV (2007) Pressure statistics and their scaling in high-reynolds-number turbulent boundary layers. J Fluid Mech 585:1–40
39. Underberg J, Cannon C, Larrey D, Makris L, Schwamlein C, Phillips H, Bloeden L, Blom D (2016) Global real-world data on the use of lomitapide in treating homozygous familial hypercholesterolemia: the lomitapide observational worldwide evaluation registry (lower), two-year data. Circulation 134(suppl_1):A12117–A12117
40. Vaghefi NS, Madnia CK (2015) Local flow topology and velocity gradient invariants in compressible turbulent mixing layer. J Fluid Mech 774:67–94
41. Warhaft Z (2000) Passive scalars in turbulent flows. Annu Rev Fluid Mech 32(1):203–240
42. Watanabe T, Sakai Y, Nagata K, Ito Y, Hayase T (2014a) Reactive scalar field near the turbulent/non-turbulent interface in a planar jet with a second-order chemical reaction. Phys Fluids 26(10):105111
43. Watanabe T, Sakai Y, Nagata K, Ito Y, Hayase T (2014b) Vortex stretching and compression near the turbulent/nonturbulent interface in a planar jet. J Fluid Mech 758:754–785
44. Watanabe T, Naito T, Sakai Y, Nagata K, Ito Y (2015) Mixing and chemical reaction at high schmidt number near turbulent/nonturbulent interface in planar liquid jet. Phys Fluids 27(3):035114
45. Watanabe T, Jaulino R, Taveira RR, da Silva CB, Nagata K, Sakai Y (2017a) Role of an isolated eddy near the turbulent/non-turbulent interface layer. Phys Rev Fluids 2(9):094607
46. Watanabe T, da Silva CB, Nagata K, Sakai Y (2017b) Geometrical aspects of turbulent/non-turbulent interfaces with and without mean shear. Phys Fluids 29(8):085105

47. Westerweel J, Fukushima C, Pedersen JM, Hunt JCR (2005) Mechanics of the turbulent-nonturbulent interface of a jet. Phys Rev Lett 95(17):174501
48. Zhuang Y, Tan H, Huang H, Liu Y, Zhang Y (2018) Fractal characteristics of turbulent–non-turbulent interface in supersonic turbulent boundary layers. J Fluid Mech 843
49. Zhuang Y, Tan H, Wang W, Li X, Guo Y (2019) Fractal features of turbulent/non-turbulent interface in a shock wave/turbulent boundary-layer interaction flow. J Fluid Mech 869

Frontiers in Social Data Domain

Efficient Text Autocompletion for Online Services

Sheng Hu

Abstract Query autocompletion (QAC) is an important interactive feature that assists users in formulating queries and saving keystrokes. Due to the convenience it brings to users, it has been adopted in many applications, such as Web search engines, integrated development environments (IDEs), and mobile devices. In my previous works, I studied several fundamental problems of QAC and developed novel QAC techniques that deliver high-quality suggestions in an efficient way. The remarkable contribution is the proposal of a novel QAC paradigm through which users may abbreviate keywords by prefixes and do not have to explicitly separate them. Another contribution is to efficiently solve geographical location constraints such as considering Euclidean distances to different locations when completing text queries. Based on the above studies, an overview of novel QAC methods across different application domains is provided in this chapter. By creating a data circulation on various QAC applications, I believe that the proposed methods are practical and easy to use in many real-world scenarios. I illustrate the realized data circulation in Sect. 2. Contributions to the society are presented in Sect. 3.

1 Introduction

In recent years, the rapid boost of massive data volumes has led to the increasing need for efficient and effective search activities. One common search activity often requires the user to submit a query to a search engine and receive answers as a list of documents in ranked order. Such a search task is called document retrieval and has been studied by the information retrieval community for many years. Before receiving the ranked documents, a query is issued to the search system. Then the documents are sorted by the relevance calculated according to the query. Query formulation

S. Hu (✉)
Hokkaido University, WPI-ICReDD, Kita 21, Nishi 10 Kita-Ku, Sapporo 001-0021, Japan
e-mail: hu.sheng@icredd.hokudai.ac.jp

K. Takeda et al. (eds.), *Frontiers of Digital Transformation*,
https://doi.org/10.1007/978-981-15-1358-9_11

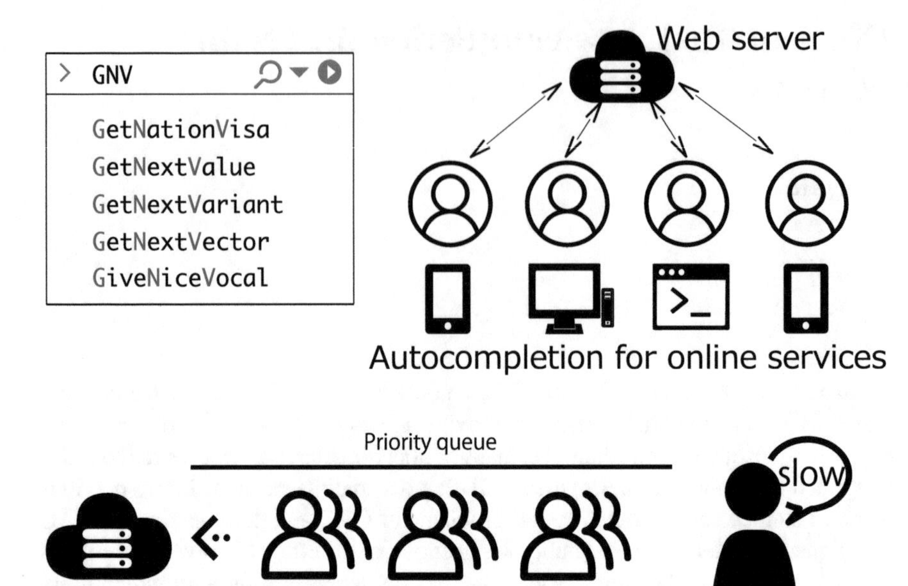

Fig. 1 Query autocompletion service

thus arose and focused on improving the overall quality of the document ranking with regard to the submitted query. As an important subtask of query formulation, query autocompletion (QAC) aims at helping the user formulate his/her query while typing only the prefix, e.g., several characters [2]. The main purpose of QAC is to predict the user's intended query and thereby save keystrokes. QAC has become a main feature of today's modern search engines, e.g., Google,[1] Bing[2] and Yahoo.[3]

In some cases, it is possible to pre-compute and index all the completion candidates (i.e., keywords or phrases) in an efficient data structure. However, when query autocompletion is applied in some complex applications, such as a Web mapping service, input method editors (IMEs), and programming integrated development environments (IDEs), keeping fast response time and accurate performance becomes challenging. This has created a need for efficient and effective *autcompletion for online services*.

In Fig. 1, I show an example of autocompletion for Web search engines. When a prefix is input, it will give a candidate list to show all the completion strings. When large amount of users are intending to obtain completions from the server, they have to wait in a priority queue thus incurring response time lags.

[1]https://www.google.com/.

[2]https://www.bing.com/.

[3]https://search.yahoo.com/.

Following this trend, the database and software engineering community has begun to investigate autocompletion for various applications recently. Some research works have been done on autocompletion for Web mappings [9, 14, 16, 17], and several autocompletion systems for IDEs have been proposed [1, 4, 10–12].

The general approach to achieving efficient and effective autocompletion is to use an index structure called trie [3], which can be taken as a search tree for fast lookups. For effective autocompletion, a straightforward way is to collect popularity statistics for each candidate and rank it according to this criterion.

In my previous works [6–8], I focused on both efficient and effective autocompletion methods. In other words, I use novel index structures to improve search performance and also collect various features for more accurate ranking. Under this setting, I combine these two ideas together to design algorithms and data structures. I focus my effort on applications for Web mappings, IMEs, IDEs, and desktop search.

In this section, I will give a brief introduction about the fundamental problems I had addressed. Sect 1.1 clarifies the problem *location-aware autocompletion*. Sect 1.2 defines the problem *autocompletion for prefix-abbreviated input*. Then, Sect 1.3 describes the problem *code autocompletion*.

1.1 Location-Aware Autocompletion

In my first work [6], I considered *location-aware query autocompletion*, one of the most important applications for Web mapping service. A query for such a service includes a prefix and a geographical location. It retrieves all the objects that are near the given location and begin with the given prefix. The location can either be a point or a range. In practice, this query can help a user to find a proper point of interest (POI) with the prefix and location provided quickly. An example is shown in Fig. 2.

Technical Details

This work is motivated by the prevalence of autocompletion in Web mapping services. In this work, I adopt two kinds of query settings: range queries and top-k queries. Both queries contain two kinds of information. One is a location, such as a point or an area. The other one is a string prefix. A point of interest will be returned if it is close to the spatial location and its textual contents begin with the given prefix.

Although there have been several solutions [9, 14, 16, 17] to location-aware query autocompletion that are based on a combination of spatial and textual indices to process queries, all of them suffer from inefficiency when the dataset is large or when large amount of simultaneous queries occur. Most existing works can be classified into text-first, space-first, and tightly combined methods, according to how

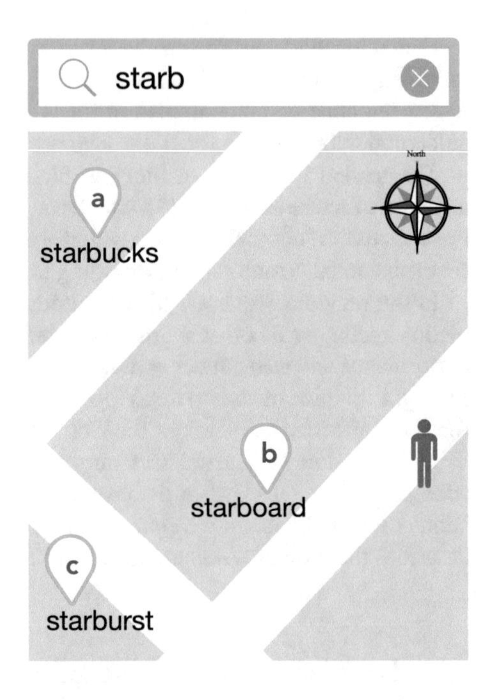

Fig. 2 Location-aware autocompletion. When a user issues "starb" as a prefix and selects the screen as the range, the system will return "starbucks", "starboard", "starburst" and display the their positions on the map

the indices are combined. The text-first methods first index the text descriptions and then apply spatial constraints as filters to verify the objects. For example, if a user searches for "Starbucks" around "New York", text-first methods will first find all the objects matching "Starbucks" and then verify whether they are around "New York". The space-first methods adopt the reverse order. The tightly combined methods will transfer "Starbucks" and "New York" into one combined token and then use it as a key for lookups. After analyzing the characteristics of queries, a novel text-first indexing method was devised to answer both range and top-k queries. I chose to store the spatial information of data objects in the trie index for fast lookups. Several pruning techniques were developed to avoid unnecessary access using pointers to quickly locate data objects. Due to these techniques, only a small part of objects need to be checked. I also extended my method to deal with error-tolerant problems which can suggest correct queries when there are typos in the user input.

Extensions

In the current work, the locations are independent points with static popularity scores on the map. For future work, I plan to make use of the boosted LBSN (Location-based Social Network) to enhance the location-location relationship for semantic autocompeltion. This allow us to take advantage of the rich properties from LBSN such as functional regions, time, user attributes, and even descriptive landscape pictures. I also plan to utilize external resources such as knowledge graphs or corpora to enrich the utilities of the methods. Moreover, I also consider multimedia techniques to combine landscape descriptions in the methods.

1.2 Autocompletion for Prefix-Abbreviated Input

My second work [8] studies *prefix-abbreviated query*, which has a wide usage in IMEs, IDEs, and desktop search. This query can retrieve relevant results when the query is a prefix-like abbreviation. I focused on both efficiency and effectiveness of this approach. An application example is to find the candidates by inputting characters as few as possible.

Technical Details

In this work, I studied a novel QAC paradigm. I proposed a novel QAC paradigm through which users may abbreviate keywords by prefixes and do not have to explicitly separate them. There is no previous work on such an implicit autocompletion paradigm. Moreover, the naïve approach of performing iterative search on a trie is also computationally expensive for the problem in this work. Thus a novel indexing and query processing scheme called nested trie is proposed to efficiently complete queries. Moreover, to suggest meaningful results, I devised a ranking method based on a Gaussian mixture model (GMM) [13], by considering the way in which users abbreviate keywords, as opposed to the traditional ranking method that merely considers popularity. Efficient top-k query processing techniques are developed on top of the new index structure. The proposed approaches can efficiently and effectively suggest completion results under the new problem setting.

Extensions

In the current work, I use a generative probabilistic model of GMM to rank the completions. In the future, I consider improving this ranking module with more complex ranking models by utilizing learning to rank techniques. I also plan to incorporate user profile information to personalize the abbreviated patterns for more accurate suggestions. Moreover, I am going to deploy the algorithms as a practical service for demonstrations.

1.3 Code Autocompletion

My third work [7] explored *scope-aware code completion*. Given a free acronym-like input of a programming method name, the proposed method can suggest the completions in a fast and accurate way. This is extremely crucial when online IDEs are becoming popular these days. The proposed method utilized the discriminative model to measure the relevance between input and different completion candidates.

Technical Details

The third work takes a different view and focuses on the ranking performance of code completion, which is a traditional popular feature for API access in IDEs. The prompted suggestions not only free programmers from remembering specific details about an API but also save keystrokes and correct typographical errors. Existing methods for code completion usually suggest APIs based on popularity in code bases or language models on the scope context. However, they neglect the fact that the user's input habit is also useful for ranking, as the underlying patterns can be used to improve the accuracy for predictions of intended APIs. In this work, I proposed a novel method to improve the quality of code completion by incorporating the users' acronym-like input conventions and the APIs' scope context into a discriminative model. The weights in the discriminative model are learned using a support vector machine (SVM) [15]. A trie index was employed to store the scope context information. An efficient top-k suggestion algorithm is developed.

Extensions

In the current work, I adapt my ranking model to all the code bases including different projects. Thus, in the future, we can consider collecting the project topic information and programmer information for more customized code completion. Furthermore, I will study the naturalness and conventions underlying in different programming languages to provide more accurate completions on top of specific programming languages. I am going to develop these techniques as plug-ins in online IDEs. Moreover, my acronym-like input paradigm can be extended as various applications to improve its usability in the real world. We can adapt the method to complete a sentence, a road network sequence, or equip it with a general speech recognition interface to improve productivities.

1.4 Summary

In the above QAC studies, I observed that the character input from users is used to produce completed candidates while the selections on different candidates are recorded in the query logs as output. Interestingly, such query logs will be learned and reused for providing improved and more precise autocompletions for new users. This forms the basis of the data circulation.

With the achievements from the above QAC studies, I realize a real-world data circulation formally. Such a circulation consists of three steps: Acquisition, Analysis, and Implementation. Acquisition here represents the observations of entire users' activities that occur from the user's first keystroke in the search box to the last selection of an autocompletion candidate provided. In the Analysis step, I focused on revealing the input behavior patterns of users by employing machine learning and statistical modeling techniques for accurate predictions. In the Implementation step, I demonstrated superior performances through prototype autocompletion systems. Details will be discussed in Sect. 2.

I also show such a data circulation can benefit the society in terms of e-commerce search systems, location-based services, and online text editors. More details can be found in Sect. 3.

2 Realized Data Circulation

First, I show an overview of the data circulation realized in my works in Fig. 3.

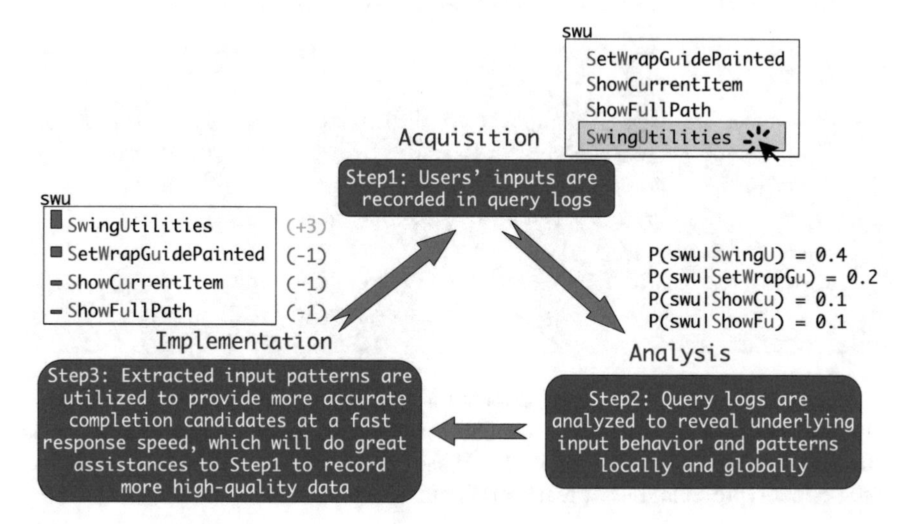

Fig. 3 Real-world data circulation in autocompletion

As Fig. 3 shows, I divide the real-world data circulation into three steps:

- Step1: Acquisition
- Step2: Analysis
- Step3: Implementation

In the general sense, Acquisition means revealing the users' real intentions or desires from the digital footprints in existing services left by the users. Analysis indicates employing the cutting-edge data analysis techniques such as statistical models, machine learning, and deep learning models to describe the underlying patterns from the observations of Acquisition. Analysis also includes utilizing novel database techniques such as indexing and query processing paradigm designs to handle scalable data volumes collected in Acquisition. A well-formulated data model will fit the users' desires properly and thus allows the provision of accurate predictions. A scalable and robust data model can handle simultaneous query requests to satisfy real-time requirements. In Implementation step, the models and prototypes obtained in Analysis are applied to specific applications. Parameters are adjusted to fit the problem settings in real-world scenarios. By inspirations and fancy ideas, innovative products and services are expected to be developed in this step.

In this section, I will describe the details of the data circulation realized in terms of autocompletion. Acquisition here represents the observations of entire users' activities that occur from the user's first keystroke in the search box to the last selection of an autocompletion candidate provided. The specific input of a user can reflect his/her input behavior and patterns. The cursor hovering time over a completion can be seemed as an implicit feedback. The final clickthrough link can be considered highly related to the user's desires. All of such information is recorded into query logs for further analysis.

For Acquisition, I have used large datasets in all my previous works [6–8].

For example:

- FSQ [6] is a dataset containing 1M POIs collected from real-world check-in data on a social media service Foursquare.[4]
- PINYIN [8] is the Chinese input method corpus compiled by collecting frequent words appearing in real-world communications. It contains 3.55M Chinese phrases.
- Java Corpus [7] is a large-scale code base collected from the real-world code bases on GitHub. It contains 10M lines of codes.

Moreover, corresponding hand-crafted training sets are collected from Amazon Mechanical Turks. I managed to handle these "Big Data" to produce efficient and effective completions in my thesis [5]. These "Big Data" are sufficient reflections and perfect representations of real-world data.

[3]https://foursquare.com/.

In the Analysis step, I focused on revealing the input behavior patterns of users by employing machine learning and statistical modeling techniques for accurate predictions. In paper [8], I made use of Gaussian mixture model (GMM) to evaluate the probability (density function) of a specific input pattern. In paper [7], I utilized the noisy channel model, logistic regression, and support vector machine (SVM) to accurately predict the user's real intended completions. I considered the input patterns and usage counts as global features as well as the scope context information as a local feature. Thus we can explore the underlying input patterns both locally and globally. A significant observation is that common input patterns of human writing behavior can evidently improve the completion accuracy. For example, users tend to preserve consonant characters more than vowel characters when inputting abbreviations. Moreover, to handle scalable query throughputs, I proposed novel indexing techniques in [6, 8].

In the Implementation step, I demonstrated superior performances through prototype autocompletion systems. The work developed in [6] can be directly applied to many location-based service (LBS) applications such as map applications, POI recommendation systems or trip route recommendation systems. My prototype system in [6] can handle a hundred thousand queries per second on a very large dataset containing 13M records. My works in [7, 8] can be directly applied to online integrated development environments (IDEs), cloud input method editors (IMEs), desktop search systems, and search engines. Experiments in [8] show that my prototype system can save on average 21.6% keystrokes at most and achieve 121 times speedup at most than the runner-up method. Experiments in [7] show that my experimental system can achieve 6.5% improvement at top-1 accuracy and achieve 31 times speedup than the naïve method. Sufficient experimental results prove that my prototype systems are superior in terms of both effectiveness and efficiency, which can be easily adapted to innovative products and services.

To show the real-world applicability of my algorithms, I incorporated the algorithms in [6] into a real-world mapping application. This application is called Loquat (Location-aware query autocompletion). I used two preprocessed datasets UK and US here. UK is a dataset containing POIs (e.g., banks and cinemas) in the UK.[5] It contains 181,549 POIs in total and the geographical coordinates are within the territory of the UK. US is a dataset containing about 2M POIs located in the US.[6]

Preprocessing: By observing that the range queries issued by users are usually located in the surroundings of large cities, I removed POIs that are too far from the metropolis. After this preprocessing, the statistics of UK and US are shown in Table 1.

[5]https://www.pocketgpsworld.com/.

[6]https://www.geonames.org/.

Table 1 Dataset statistics

Dataset	# of POIs (raw data)	# of POIs (after removal)
UK	181,549	122,748
US	2,234,061	1,782,024

Fig. 4 Range Query

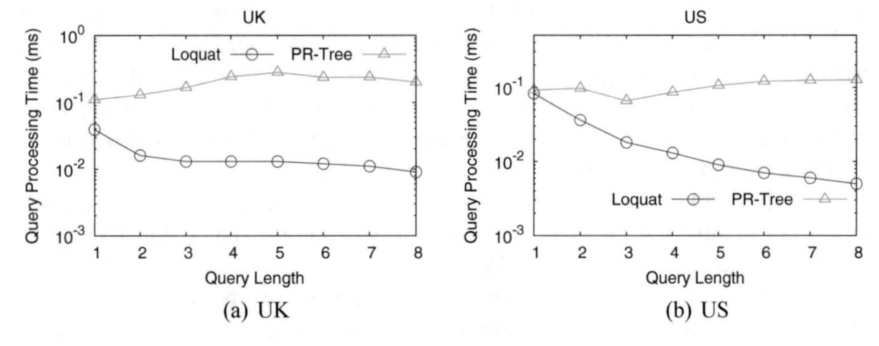

Fig. 5 Top-k Query

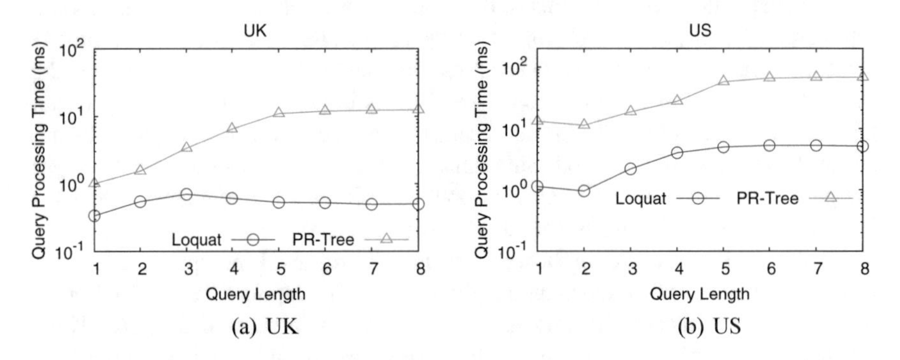

Fig. 6 Range Query with Error = 3

Here I will show the query processing time of my system in comparison to the state-of-the-art method named Prefix-Region Tree [17] (denoted as PR-Tree).

Figure 4a, b show the efficiency on range queries on UK and US, respectively. The proposed Loquat is at most 30 times faster than PR-Tree on the smaller dataset UK and at most 10 times faster than PR-Tree on the larger dataset US.

Meanwhile, Fig. 5a, b shows the efficiency on top-k queries on UK and US, respectively. The proposed Loquat is at most 20 times faster than PR-Tree on both UK and US datasets.

Figure 6a, b shows the efficiency on range queries with $error = 3$ (i.e., tolerating three typos) on UK and US, respectively. The number of errors is set to three because

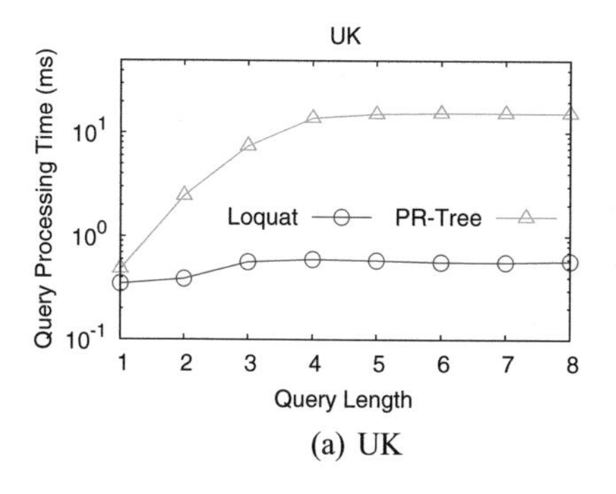

(a) UK

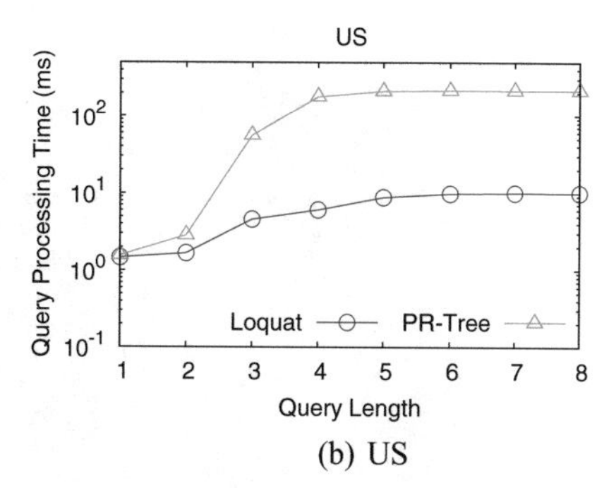

(b) US

Fig. 7 Top-k Query with Error = 3

existing works [3, 15] proved that in most real-world scenarios, the number of typos is less than three. The proposed Loquat is at most 25 times faster than PR-Tree on UK and 13 times faster on US datasets.

Meanwhile, Fig. 7a, b shows the efficiency on range queries with $error = 3$ on UK and US, respectively. The proposed Loquat is at most 30 times faster than PR-Tree on UK and 22 times faster on US.

High-performance products and services in Implementation will also help to provide more high-quality data for the collection in Acquisition, which forms an occlusive circulation in terms of autocompletion.

Figure 3 gives an example to illustrate the data circulation of the work in [7]. First begins with the Acquisition step. Suppose a user types an input "swu", then all the matched candidates are listed in a naïve alphabetical order as `SetWarpGuide Painted`, `ShowCurrentItem`, `ShowFullPath`, `SwingUtilities`. After that, let's say that the user clicked `SwingUtilities` as his/her intended completion. Then the Analysis step begins. Based on the observations in Acquisition, the machine learning model is trained to let it have a high probability to match "swu" with "SwingU" but low probabilities with the others. Note that P(Q|C) in Fig. 3 represents the probability of abbreviating C as Q if the user's intended completion is C. Last is the Implementation step. After training the model, `SwingUtilities` will have a higher rank (move up by 3 positions) in the autocompletion candidates. Thus we can obtain more accurate predictions by learning the user input patterns.

3 Contributions to the Society

Autocompletion can provide great assistance in human knowledge explorations. The data circulation formed in my works can help better understand the search intentions of users, which can benefit the society from many aspects.

First, in traditional search systems under e-commerce settings, an effective query autocompletion can directly lead to a purchase decision. Figure 8 shows an example

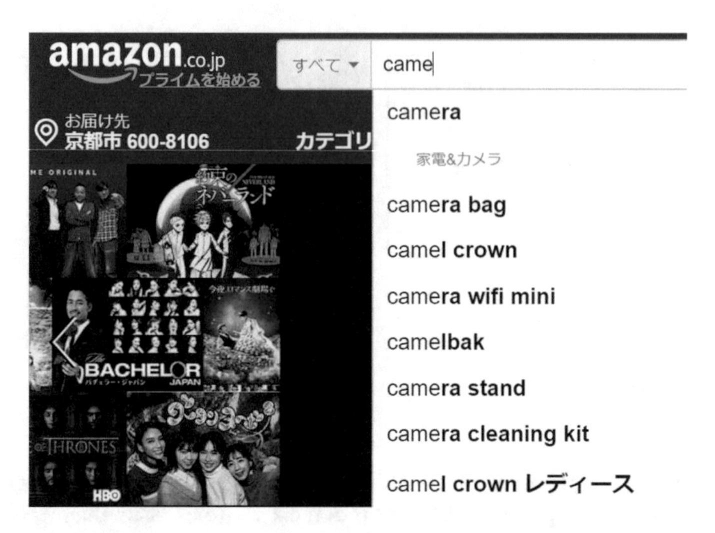

Fig. 8 Autocompletion in Amazon

of the autocompletion system of Amazon.[7] A customer-personalized autocompletion system can stimulate consumer spending and save the time of users, which might boost the economic growth of a country as well as improve customer satisfaction.

Second, an effective and efficient autocompletion system for Location-based service can promote the local tourism development. Figure 9 shows an example to autocomplete the query as "Restaurants near me". Comparing with existing systems such as Google Maps[8] which can only tolerate a few typos in its autocompletion, my work [6] can tolerate more typos in a dynamic way. This is extremely important when the location names are too long, e.g., "Gasselterboerveenschemond", the name of a village located in the Netherlands. A personalized autocompletion system can balance the user preferences and spatial distances to provide proper rankings. A scalable and robust autocompletion system can support large query throughputs and thus provide smooth and stable services. Therefore, it will help to increase the venues of restaurants, museums, hotels and tourist attractions.

Third, a quick-response online IDE editor with accurate API suggestions can help the programmers or designers to improve their productivities in their work as shown in the example in Fig. 10. With the growing popularity of online text editors / IDEs (e.g., Overleaf[9] and IBM Bluemix[10]), the demand on efficiency and effectiveness is increasing. Comparing with these existing online IDEs, my works [7, 8] can support acronym-like input and thus save almost 30% keystrokes. Especially

[7]https://www.amazon.co.jp/.

[8]https://www.google.com/maps/.

[9]https://overleaf.com/.

[10]https://www.ibm.com/cloud.

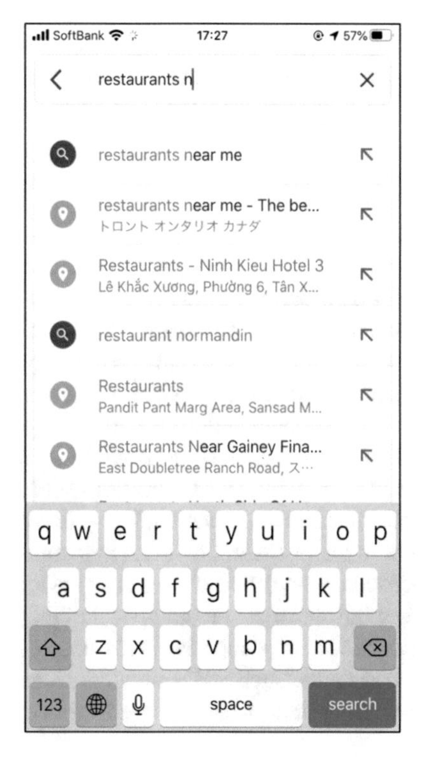

Fig. 9 Autocompletion in Goolge Maps

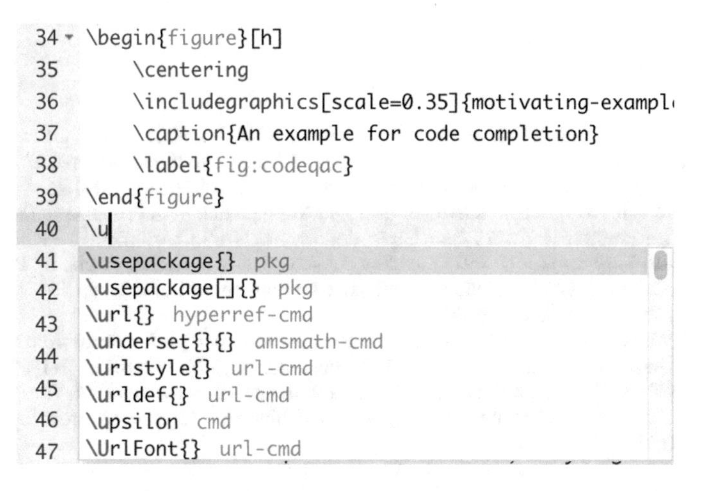
```
34 \begin{figure}[h]
35     \centering
36     \includegraphics[scale=0.35]{motivating-exampl
37     \caption{An example for code completion}
38     \label{fig:codeqac}
39 \end{figure}
40 \u
41 \usepackage{}  pkg
42 \usepackage[]{}  pkg
43 \url{}  hyperref-cmd
44 \underset{}{}  amsmath-cmd
45 \urlstyle{}  url-cmd
46 \urldef{}  url-cmd
47 \upsilon  cmd
   \UrlFont{}  url-cmd
```

Fig. 10 Autocompletion in online IDEs

in teamwork, an effort-saving and well-performed code completion system will boost the coordination among multiple workers, which will definitely create more value for the whole society.

4 Conclusion

In this chapter, several fundamental autocompletion problems were investigated for online services in a wide variety of fields in recent decades. There are many practical challenges for online real-time autocompletion in the real world. Particularly, efficiency and effectiveness challenges are the most important ones which will influence the response time and prediction accuracy of an autocompletion system directly. Hence, a real-world data circulation is established to address those challenges. Such a circulation consists of three steps: Acquisition, Analysis, and Implementation. The Acquisition here represents the observations of entire users' activities that occur from the user's first keystroke in the search box to the last selection of an autocompletion candidate. In the Analysis step, I focused on revealing the input behavior patterns of users by employing machine learning and statistical modeling techniques for more accurate predictions. In the Implementation step, superior performances were shown using the prototype autocompletion systems. Such a data circulation can benefit our society in terms of e-commerce search system, location-based service, and online text editors.

References

1. Asaduzzaman M, Roy CK, Schneider KA, Hou D (2014) CSCC: simple, efficient, context sensitive code completion. In: Proceedings 2014 IEEE international conference on software maintenance and evolution, pp 71–80. https://doi.org/10.1109/ICSME.2014.29
2. Cai F, de Rijke M (2016) A survey of query auto completion in information retrieval. Found Trends Inform Retr 10(4):273–363. https://doi.org/10.1561/1500000055
3. Chaudhuri S, Kaushik R (2009) Extending autocompletion to tolerate errors. In: Proceedings 2009 ACM SIGMOD international conference on management of data, pp 707–718, https://doi.org/10.1145/1559845.1559919
4. Han S, Wallace DR, Miller RC (2011) Code completion of multiple keywords from abbreviated input. Autom Softw Eng 18(3–4):363–398. https://doi.org/10.1007/s10515-011-0083-2
5. Hu S (2019) Efficient text autocompletion for online services. Doctoral Thesis at Graduate School of Information Science, Nagoya University. https://www.irdbniiacjp/01152/0004293757
6. Hu S, Xiao C, Ishikawa Y (2018) An efficient algorithm for location-aware query autocompletion. IEICE Trans Inf Sys 101-D(1):181–192. http://search.ieice.org/bin/summary.php?id=e101-d_1_181
7. Hu S, Xiao C, Ishikawa Y (2019a) Scope-aware code completion with discriminative modeling. J Inf Proc 27:469–478. https://doi.org/10.2197/ipsjjip.27.469
8. Hu S, Xiao C, Qin J, Ishikawa Y, Ma Q (2019b) Autocompletion for prefix-abbreviated input. In: Proceedings 2019 ACM SIGMOD international conference on management of data, pp 211–228. https://doi.org/10.1145/3299869.3319858

9. Ji S, Li C (2011) Location-based instant search. In: Proceedings 2011 international conference on scientific and statistical database management, pp 17–36. https://doi.org/10.1007/978-3-642-22351-8_2
10. Nguyen AT, Nguyen TN (2015) Graph-based statistical language model for code. In: Proc. 37th international conference on software engineering (ICSE), pp 858–868. https://doi.org/10.1109/ICSE.2015.336
11. Nguyen AT, Nguyen TT, Nguyen HA, Tamrawi A, Nguyen HV, Al-Kofahi JM, Nguyen TN (2012) Graph-based pattern-oriented, context-sensitive source code completion. In: Proceedings 34th international conference on software engineering (ICSE), pp 69–79. https://doi.org/10.1109/ICSE.2012.6227205
12. Nguyen TT, Nguyen AT, Nguyen HA, Nguyen TN (2013) A statistical semantic language model for source code. In: Proceedings 9th joint meeting on foundations of software engineering, pp 532–542. https://doi.org/10.1145/2491411.2491458
13. Reynolds DA (2009, Springer) Gaussian mixture models. Encyclopedia of Biometrics 741
14. Roy SB, Chakrabarti K (2011) Location-aware type ahead search on spatial databases: Semantics and efficiency. In: Proceedings 2011 ACM SIGMOD international conference on management of data, pp 361–372. https://doi.org/10.1145/1989323.1989362
15. Schaback J, Li F (2007) Multi-level feature extraction for spelling correction. In: Proceedings IJCAI-2007 workshop on analytics for noisy unstructured text data, pp 79–86
16. Zheng Y, Bao Z, Shou L, Tung AKH (2015) INSPIRE: a framework for incremental spatial prefix query relaxation. IEEE Trans Knowl Data Eng 27(7):1949–1963. https://doi.org/10.1109/TKDE.2015.2391107
17. Zhong R, Fan J, Li G, Tan K, Zhou L (2012) Location-aware instant search. In: Proceedings 21st ACM international conference on information and knowledge management, pp 385–394. https://doi.org/10.1145/2396761.2396812

Coordination Analysis and Term Correction for Statutory Sentences Using Machine Learning

Takahiro Yamakoshi

Abstract Statutes form an essential infrastructure that sustains and improves human society. They are continuously updated as society changes; that is, legislation (interpreting, drafting, and managing statutes) is an endless work. In Japan, statutes are written following a number of rules regarding document structures, orthography, and phraseology. Therefore, we need to pay attention to the formalism in addition to the substantial contents in drafting a statute. Another issue in Japanese statutes is complex hierarchical coordination that impedes smooth comprehension. In this chapter, machine-learning-aided methodologies for interpreting support and drafting support of Japanese statutory sentences are discussed. As for interpreting support, a coordination analysis methodology that utilizes neural language models is established. By transforming the whole sentences into vectors, the proposed methodology takes broader contexts into account in judgment. As for drafting support, a legal term correction methodology that finds misused legal terms and offers their correction ideas is established. Classifiers for legal term correction are utilized by regarding this task as a kind of sentence completion test with choices. Legislation itself can be regarded as a real-world data circulation of statutes from the viewpoint that legislation is to recognize problems in the real world and to solve them by updating statuses. Both coordination analysis and legal term correction support this data circulation by accelerating interpretation and drafting of statutes, respectively. Furthermore, each technology constructs another data circulation centered on its analysis results.

1 Statutory Sentence and Its Characteristics

Laws are the fundamental components of society. They illumine the exemplary ways of economic and social activities, which guarantee the health and cultural lives of citizens. Laws are continuously updated in accordance with the changes of economic situations, values, and technology.

T. Yamakoshi (✉)
Nagoya University, Nagoya, Japan

K. Takeda et al. (eds.), *Frontiers of Digital Transformation*,
https://doi.org/10.1007/978-981-15-1358-9_12

Statutes are laws written as documents, which are continuously revised in accordance with societal changes. Since statutes bind people by rights and duties, statutes must be described strictly so that they can be correctly understood. Statutes must be completely devoid of errors or inconsistencies. To meet this requirement, the language with which statutes (*statutory sentences*) are composed often contain specific wordings, technical terms, and sentence structures that are rarely used in daily written language. Overall, for understanding statutes, large amounts of knowledge and experience are required about such description rules of statutory sentences.

Japan's current statutory law system, which was established in the Meiji era,[1] has basically remained unchanged for over 100 years. Japan established its legislative (interpreting, producing, and updating statutes) technique for the precise and uniform drafting, enacting, and modifying of statutes. This technique includes a number of concrete customs and rules (hereinafter *legislation rules*) stipulated by various workbooks [1, 2].

In its first layer, the legislation rules stipulate that a statute structure must be defined as a document. Here are some examples of such rules. A statute is required to have a title and must be divided into main and supplementary provisions. Main provisions should generally be divided into articles.

Along with statute structure definitions, the technique enacts rules for forming, modifying, and deleting particular sentences, figures, and tables. In addition to such editing guidelines, rules exist on the usage of *kanji* and *kana*. Furthermore, they define the distinct usage of similar legal terms. For example, three Japanese words, "者 (a)," "物 (b)," and "もの (c)," are all pronounced *mono* and share the concept of "object." Nontheless, term (a) only indicates a natural or a juristic person, term (b) only indicates a tangible object that is not a natural or a juristic person, and term (c) only indicates an abstract object or a complex of such objects. Figure 1 displays phrases including these legal terms. In ordinary Japanese written language, unlike in statutory sentences, (c) can refer to "者" and "物" . That is, we can use (c) as "著作物" (work) を (accusative marker) 創作する (create) <u>もの</u> (*momo*) to express "a person who creates a work" in ordinary Japanese. However, this usage is prohibited in Japanese statutory sentences due to the above rules.

Two more examples of such legal terms are "及び (d)" (*oyobi*) and "並びに (e)" (*narabini*), which are coordinate conjunctions indicating "and" coordination. The legislation rules provide special directives for these legal terms to express the *hierarchy* of coordination. Term (d) is used for coordinations of the smallest hierarchy, and term (e) is used for the coordinations of the non-smallest hierarchy. For example, the sentence in Fig. 2 contains two coordinate structures of (d) and (e) that compose a two-layered hierarchical coordinate structure.

These legislation rules are not only applicable for the central government. Local governments enact ordinances and orders using identical legislation rules. Further-

[1] Japan established its law system by referencing Occidental countries' law systems in this era after its opening to the world in the late 19th century.

Fig. 1 Phrases in Japanese statutory sentences with legal term (underlined), from the Copyright (Act No. 48 of 1970)

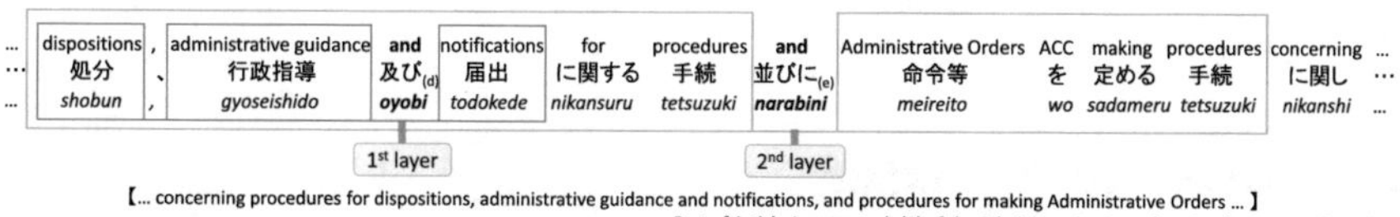

Fig. 2 Japanese statutory sentence with "*oyobi*$_{(d)}$" and "*narabini*$_{(e)}$"

more, such private entities as companies also use the rules to enforce contracts, agreements, articles of incorporation, patents, and other legal documents. Mastering legislation rules is critical for those who handle any kind of legal documents.

Under this context of Japanese statutory sentences, we identify two issues for handling them. The first issue is the strong influence of legislation rules on drafting. These legislation rules are comprehensive. Legislation bureaus scrutinize draft statutory sentences, which force legislation officers to completely obey the legislation rules in drafting a statute. That is, they need to be cognizant of the formalism of the statutory sentences in addition to their substantial contents.

Another issue is the appearance of long, complex statutory sentences originated by hierarchical coordinate structure, which complicates their understanding. Those who attempt to read such long and complex statutory sentences are required to have sufficient experience and knowledge reading them or else they will most likely fail to identify and understand their sentence structures. Machines that handle such sentences will also probably fail to adequately analyze their sentence structures because of their complexity, and analysis errors will cause deterioration of further processes.

2 Solutions for the Issues

We establish a solution for each issue. For the issue of the hierarchical coordinate structure, we propose a novel coordination analysis method specialized for Japanese statutory sentences. With coordination analysis, we can simplify long and complex statutory sentences, which supports any person or system that has trouble understanding long and complex statutory sentences. Thus, this study can be regarded as a quite fundamental study for further sentence processes.

For the issue of legislation rules on drafting, we establish a legal term correction methodology as a first step toward establishing the proofreading of comprehensive statutory sentences. Our legal term correction finds misused legal terms and offers correction ideas for them, which helps legislation officers to write statutory sentences consistent with the legislation rules.

We briefly describe the background, the method, and the experimental result of each solution in the following sections.

2.1 Coordination Analysis

Background

Japanese statutory sentences are not easy to read because of the frequent appearance of hierarchical coordinate structures that is supported by the Japanese legislation rules. Various methods have been proposed for coordination analysis for general sentences (e.g., [3–8]). However, we cannot expect these methods, especially those for general sentences, to work well on Japanese statutory sentences because they are not designed to consider the legislation rules. In fact, even a famous Japanese text parser, KNP (v3.01) [5], could identify coordinate structures in Japanese statutory sentences at only 26 points of F-measure [9].

One existing method specialized for Japanese statutory sentences is from Matsuyama et al. [9]. This method deterministically identifies the hierarchy of coordinate structures and the scope of their conjuncts based on the legislation rules. However, this method has a weak point in identifying conjuncts whose word length is very different from that of their adjacent conjuncts. This is because this method uses a scoring strategy based on one-to-one word alignment in identifying conjuncts. Also, this word alignment approach does not consider context outside the coordinate structure, which may also degrade the performance.

From this background, we propose a new method for identifying hierarchical coordinate structures in Japanese statutory sentences. We use Neural Language Models (NLMs) [10], especially Long-Short Term Memory (LSTM) [11] based NLMs, to overcome the weak points of the conventional method.

Method

Analysis of hierarchical coordinate structures consists of two subtasks: *hierarchy identification* where we identify the hierarchical relationship among coordinate structures and *conjunct identification* where we identify the scopes of conjuncts in a coordinate structure.

Hierarchy Identification
As for hierarchy identification, our method inherits the deterministic analysis strategy from Matsuyama et al.'s method. This strategy consists of the following processes:

1. Coordinator Extraction and Ranking
   Our method first extracts coordinators (coordinate conjunctions, etc. that indicate coordination) from the given sentence by text-matching. It then decides the analysis order of coordinate structures according to coordinator types. The hierarchy of the coordinate structures in the sentence is roughly consolidated by this analysis order. For example, it analyzes "$oyobi_{(d)}$" in prior to "$narabini_{(e)}$" to let the latter coordinate structure include the former one.
2. Conjunct Candidate Extraction
   For each coordinator, our method extracts candidates of the conjuncts adjacent to the coordinator. The extraction is rule-based that involves part-of-speech matching. Our method also uses heuristics for hierarchy identification such as a coordinate structure of "$narabini_{(e)}$" should include at least one coordinate structure of "$oyobi_{(d)}$."
3. Conjunct Identification
   Our method chooses the most likely conjunct candidates using NLMs, which we describe later.
4. Further Conjunct Existence Judgment
   A coordinate structure may have more than two conjuncts (see the coordinate structure of "$oyobi_{(d)}$" in Fig. 2). In the first conjunct candidate extraction, this method extracts candidates for the last two conjuncts.[2] After their identification, it judges whether there is another conjunct forehand of the conjuncts in this process. If it judges that another conjunct exists, it then proceeds to the conjunct candidate extraction for the next conjunct.
5. Substitution of Coordinate Structure
   After identifying all the conjuncts in a coordinate structure, the method substitutes the coordinate structure with its last conjunct. This substitution contributes to fairer conjunct identification by flattening long coordinate structures into simple phrases.

Conjunct Identification
In contrast to hierarchy identification, our method adopts LSTM-based NLMs instead of one-to-one word alignment for conjunct identification. They transform each conjunct candidate into a fixed-length vector so that our method can identify conjunct

[2]This is always satisfied in Japanese statutory sentences because coordinators are to be located between the last two conjuncts of coordinate structures in such sentences.

scopes without being affected by their length. Since they are trained by tokenized statutory sentences, our method does not rely on any annotated dataset such as Genia Treebank [12], which is friendly for a domain with limited resources such as Japanese statutory sentences.

Our method identifies conjunct scopes based on two assumptions on coordinate structures: (1) conjunct similarity and (2) conjunct interchangeability. The conjunct similarity is that paired conjuncts are alike. Assume that we have a sentence "I ate a strawberry cake and an orange cookie for a snack today." The most likely coordinate structure in this sentence is "a strawberry cake and an orange cookie," where "a strawberry cake" and "an orange cookie" are the paired conjuncts. Both conjuncts are similar in terms of that both objects are confections with fruit. With NLMs, our method evaluates the conjunct similarity from the similarity of word distributions at the left and right edges of the two conjuncts. In the previous example, we calculate the word distributions of X in the sequences "X a strawberry cake for a snack today" and "X an orange cookie for a snack today," where "ate," "eat," "chose," etc. may have high probabilities in both of the sequences. In the same manner, we calculate the word distributions of X in "I ate a strawberry cake for a snack today X" and "I ate an orange cookie for a snack today X," where punctuations may have high probabilities in both of them. This assumption is inspired by the distributional hypothesis [13] that a word's meaning is influenced by the position where it is used.

Conjunct interchangeability is that the fluency of a sentence is maintained even if the conjuncts in a coordinate structure are swapped with each other. In the previous example, we swap the two conjuncts "a strawberry cake" and "an orange cookie" as we have a sentence "I ate an orange cookie and a strawberry cake for snack today." This swapped sentence seems fluent so that we consider this coordinate structure is adequate. We calculate the fluency by inputting the swapped sentence to the NLMs.

Experiment

In the experiment, we compared our method with Matsuyama et al.'s method. We constructed both input data and gold data from an annotated corpus [14]. In the corpus, each sentence is annotated by hand with information on morphological analysis, *bunsetsu* segmentation, and dependency analysis. We constructed the input data by removing information on dependency analysis and extracting each parenthesized expression as another independent sentence. The size of the input data is 592 sentences with 792 coordinate structures. We created the gold data using the dependency information in the corpus.

We built LSTM-based NLMs [15] for our method. Based on our preliminary experiment, we decided that each model has four hidden layers and that each layer consists of 650 LSTM units. A sequence of one-hot word vectors is fed into the model. In creating the one-hot word vector, we used basic forms of words obtained from a Japanese morphological analyzer MeCab (v0.98) with the IPA dictionary [16].

Table 1 Experimental results

		Our method	Matsuyama et al.
Coordinate structure	*P*	**66.1%**	46.8%
		(463/700)	(312/667)
	R	**64.6%**	43.5%
		(463/717)	(312/717)
	F	**65.2**	45.1
Conjunct	*P*	**83.0%**	68.0%
		(1,372/1,653)	(1,019/1,499)
	R	**81.0%**	60.2%
		(1,372/1,694)	(1,019/1,694)
	F	**82.0**	63.8

Table 1 shows the experimental results of each method. Since our method was much superior to the conventional methods in all evaluation indices, we confirmed its effectiveness for coordination analysis of Japanese statutory sentences.

2.2 *Legal Term Correction*

Background

Legislation drafting requires careful attention. To avoid errors and inconsistencies, we should inspect statutory sentences in accordance with the legislation rules. However, inspections of statutory sentences are still conducted mainly by human experts in legislation, which requires deep knowledge and an enormous amount of labor. The legislation rules are applied to not only statutes but also ordinances and orders of local governments. Furthermore, legal documents in a broad sense, such as contracts, terms of use, and articles of incorporation, are also written in compliance with the rules. Manual inspections are also dominant in these domains.

From this background, this work aims to establish a proofreading method for statutory sentences. Legal terms with distinct usage are especially focused, since judgments of such legal terms require understanding contexts around the legal terms so that a simple rule-based approach is insufficient.

Problem Definition

Although there are a lot of studies on proofreading methods [17–20], no study other than the presented work focuses on legal terms that are used separately on the basis of context. Therefore, a definition of the legal term correction task is provided below so that we can search for a solution.

Concretely, we define it as a special case of the multi-choice sentence completion test. For instance, we assume that we are going to check legal terms of the following sentence "著作者　著作物を創作するものをいう。" ("Author" means a *mono* who creates a work.) In this casx e, the underlined legal term "もの (c)" (*mono*) is to be checked, which constitutes the legal term set {者 (a), 物 (b), もの (c)}. We then blank the legal term in the sentence such as "著作者　著作物を創作する___をいう。" and try to pick the best word from the legal term set that most adequately fills the blank.

We use a scoring function score(W^l, t, W^r) to pick the best candidate, where W^l and W^r are two word sequences adjacent to the left and right of legal term t, respectively. In the above case, W^l and W^r are W^l = 著作物を創作する (*chosakubutu wo sosakusuru*; creating a work) and W^r = をいう。(*wo iu*.; means). We expect the function to assign the highest score to term (a) and to output a suggestion that "もの (c)" in the sentence be replaced into '者 (a)".

Approaches

Two classifier-based approaches are proposed here for the scoring function.

Random Forest Classifiers

With this setting, first an approach that uses Random Forest classifiers [21] is proposed, each of which is optimized for each set of similar legal terms. The classifiers input words adjacent to the targeted legal term and output the most adequate legal term in the targeted legal term set. Here, we prepare distinct sets of decision tree classifiers for each legal term set T. Concretely, the scoring function for a legal term set T using Random Forest classifiers $\mathrm{score}_{\mathrm{RF}_T}$ is defined as follows:

$$\begin{aligned} &\mathrm{score}_{\mathrm{RF}_T}(W^l, t, W^r) \\ &= \sum_{d \in D_T} P_d(t | w^l_{|W^l|-N+1}, \ldots, w^l_{|W^l|-1}, w^l_{|W^l|}, w^r_1, w^r_2, \ldots, w^r_N), \end{aligned} \tag{1}$$

where D_T is a set of decision trees for the legal term set T and $P_d(t | w^l_{|W^l|-N+1}, \ldots, w^l_{|W^l|-1}, w^l_{|W^l|}, w^r_1, w^r_2, \ldots, w^r_N)$ is the probability (actually 0 or 1) that $d \in D_T$ chooses a legal term $t \in T$ based on features $w^l_{|W^l|-N+1}, \ldots, w^l_{|W^l|-1}, w^l_{|W^l|}, w^r_1, w^r_2, \ldots, w^r_N$. w^l_i and w^r_i are the i-th word of W^l and W^r, respectively, and N is the *window size* (the number of left or right adjacent words). Here, the rightmost N words of W^l are used because they are the nearest words to t.

Although language models [22–24] are typically used for the multi-choice sentence completion test, Random Forest classifiers are adopted here for the following two reasons: First, we can optimize hyperparameters of the Random Forest classifier for each legal term set, which can cope with the characteristics of the legal term set more flexibly. Second, a classifier approach can easily cope with multi-word legal term sets. Multi-word legal terms are quite common in Japanese statutory sentences, such as "前項 の 場合 に おいて" (*zenko no baai ni oite*) and "前項 に 規定する場合 に おいて" (*zenko ni kiteisuru baai ni oite*).[3] Classifiers including Random Forest treat a legal term as one class regardless of its word count, while language models are designed to predict a single word from the given context.

BERT Classifiers
Another approach that uses a classifier based on BERT (Bidirectional Encoder Representations from Transformers) [25] is proposed here. A BERT classifier captures an abundant amount of linguistic knowledge by fine-tuning a "ready-made" model that is pretrained by a large quantity of text. Furthermore, it utilizes more contexts than the conventional classifiers in prediction, since BERT classifiers can handle whole sentences (128 tokens maximum in our experiment). The BERT classifier here inputs a "masked" sentence where the targeted legal term t is masked and outputs a probability distribution of the legal terms in t's legal term set. Therefore, this BERT classifier is a sentence-level classifier. The following equation shows the scoring function using a BERT classifier.

$$\text{score}_{\text{BERT}}(W^l, t, W^r) = \text{BERT}(t|S), \tag{2}$$

where $\text{BERT}(t|S)$ is a probability of t that the BERT classifier assigns from the masked sentence S made as follows:

$$S = \text{pp}(w_1^l w_2^l \cdots w_{|W^l|}^l \text{ [MASK] } w_1^r w_2^r \cdots w_{|W^r|}^r), \tag{3}$$

where $\text{pp}(W)$ is a function to truncate the input sentence W on the masked legal term "[MASK]" that was originally t.

Experiment

A statutory sentence corpus from e-Gov Statute Search[4] provided by the Ministry of Internal Affairs and Communications, Japan was compiled for the experiment. This corpus includes 3,983 Japanese acts and cabinet orders on May 18, 2018. Each statutory sentence in the corpus was tokenized by MeCab (v.0.996), which is a Japanese morphological analyzer. The statistics of the corpus are as follows: the total number of sentences is 1,223,084, the total number of tokens is 46,919,612, and the total number of different words is 41,470. The 3,983 acts and cabinet orders in the corpus were divided into training data and test data. The training data has 3,784 documents, where there are 1,185,424 sentences and 43,655,941 tokens in total. The test data has 199 documents with 37,660 sentences and 1,557,587 tokens

[3]Both terms mention the preceding paragraph. The difference is that the former refers to the entire paragraph, and the latter mentions the condition prescribed in the paragraph.

[4]http://elaws.e-gov.go.jp/.

Table 2 Overall performance

Classifier	acc_{micro} (%)	$acc_{macro-S}$ (%)	$acc_{macro-T}$ (%)
BERT (approach 2)	**97.57**	**96.15**	**92.56**
RF (approach 1)	95.37	93.22	84.68
TextCNN	95.99	94.12	86.28
CBOW	88.82	84.65	74.94
Skipgram	75.42	63.07	65.68
vLBL	80.23	75.46	74.17
vLBL(c)	91.38	86.32	80.67
vLBL+	90.95	85.62	81.12
Trigram	87.12	85.81	69.36
4-gram	88.81	87.83	72.58
MLE	78.61	62.49	38.81

in total. 27 legal term sets were defined by referencing the Japanese legislation manual [1, 2]. There are 7,072,599 and 251,085 legal terms in the training data and the test data, respectively.

The proposed BERT classifier and Random Forest classifiers (abbreviated as RF) were compared with the following classifiers and language models: CBOW [22], Skipgram [22], vLBL [23], vLBL(c) [24], vLBL+vLBL(c) (abbreviated as vLBL+) [24], and n-gram language. To test a neural-based model whose complexity is between BERT and the NLMs, TextCNN [26] was additionally tested, which is a sentence classifier based on a convolutional neural network.

Table 2 shows the overall performance of each model. Here, the accuracy of predicting legal terms were measured in three averages: micro average acc_{micro}, macro average by legal term set $acc_{macro-S}$, and macro average by legal term $acc_{macro-T}$.

The proposed BERT classifier achieved the best performance in all of acc_{micro}, $acc_{macro-S}$, and $acc_{macro-T}$. Notably, its $acc_{macro-T}$ is 92.56%, which is 7.88 points better than Random Forest. TextCNN achieved the second-best performance in all of the criteria as it performed better than Random Forest.

3 Real-World Data Circulation

In this chapter, we shall discuss the relationship between the work in this chapter and real-world data circulation. In Sect. 3.1, the concept of the real-world data circulation is described. In Sect. 3.2, data circulations and their contributions that can be found in this work are discussed.

3.1 Concept of Real-World Data Circulation

The real-world data circulation is a recent interdisciplinary study field on the utilization of real data. The core objective of the real-world data circulation is a sustainable improvement of the real world by circular data utilization. The circular data utilization is typically categorized into three phases of data processing: data acquisition, data analysis, and data implementation. In the data acquisition phase, we obtain live data from the real world in a computer-readable form. In the data analysis phase, we analyze the acquired data to figure out knowledge that may affect the real world favorably. In the data implementation phase, we return the knowledge to the real world, which brings about a more comfortable world. Here, it is necessary to sustain this data processing, or it cannot form a circulation.

Among many tasks where we can establish a real-world data circulation, we can consider autonomous driving as a typical example. In the data acquisition phase, for example, we identify the terrain around the car from sensors like LiDAR (Light Detection and Ranging) or GPS (Global Positioning System), where the former can directly identify the terrain by point cloud while the latter is to associate with prebuilt terrain data from a database. Using such data, we find the best trajectory for the car that avoids hitting obstacles and that brings about a smooth movement, which is the data analysis phase. Finally, we control the car autonomously according to the computation, which is the data implementation phase. By storing the terrain data from the sensors, we can continue improving our computation model using it. That is, our autonomous driving system is sustainable in terms of continuous improvement.

3.2 Examples of Real-World Data Circulation in Legislation

Data Circulation of Statutes in the Society
Fundamentally, legislation itself can be regarded as the data circulation of statutes in the real world. Figure 3 illustrates the overview of that circulation. Here two circulations are mutually related: the circulation of societal conditions and the circulation of legislation.

The circulation of societal conditions basically consists of two states: healthy and problematic. A healthy society becomes problematic when it violates the human rights of its citizens. This typically happens due to such environmental changes as technology innovation, climate changes, and economic changes. Legislation copes with the problems by updating the current statutes so that they guarantee the human rights. For example, the 2020 amendment to the Japanese Act on the Protection of Personal Information improves the scheme of HTTP cookie usage [27]. Advertisement services predict customer segments such as the age range and interests by HTTP cookies of a user gathered from Web sites he/she accessed, which has been enabled by the innovation of big data analysis. In the current scheme, cookies can be transferred between the advertisement services and Web sites without the user's

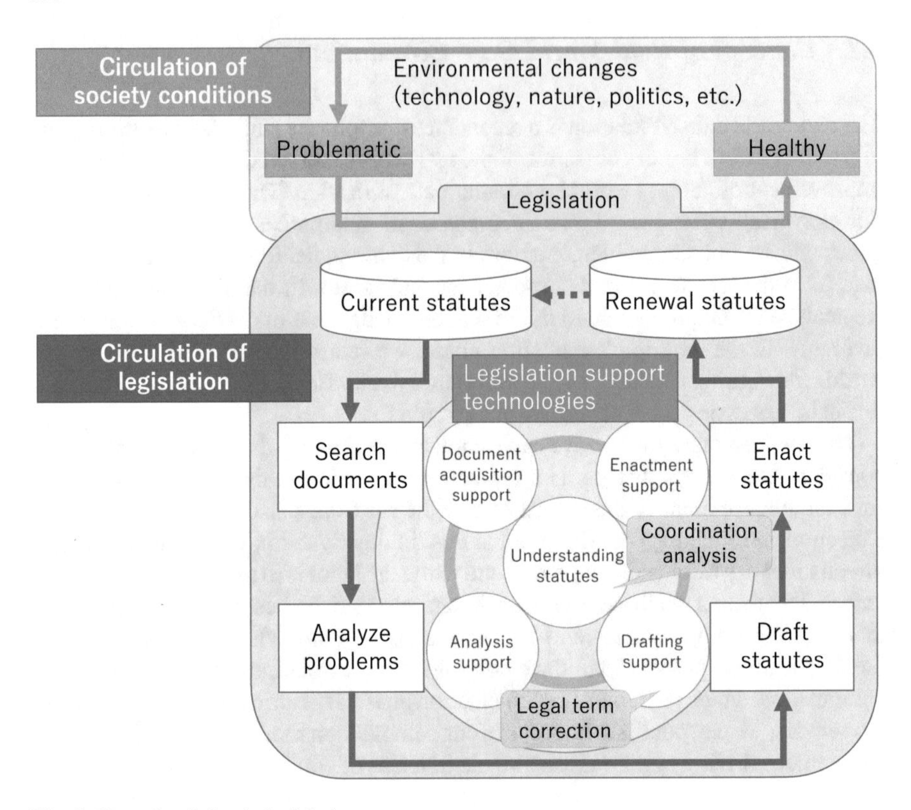

Fig. 3 Data circulation in legislation

prior consent because such data do not directly identify an individual. However, the Web site can link such data with personal data in its membership system, that is, it can acquire additional personal data without consent. To solve this vulnerability of privacy protection, the new amendment stipulates that the Web site shall find the agreement of the users on the cookie data transfer to another entity beforehand.

The circulation of legislation consists of four phases: search, analysis, drafting, and enactment. In the search phase, we thoroughly research related statutes, a phase that resembles the data acquisition phase in the context of real-world data circulation. In the analysis phase, we decide what kind of statute is needed or what statutes should be changed by analyzing the current statutes and current social situations. In the drafting phase, we verbalize the renewed statute in accordance with the legislation rules, which is one perspective of the implementation phase that focuses on documentation. In the enactment phase, we spread the revised statutes to the public, and thus people conduct economic and social activities healthily under the revised statutes. This is another perspective of the implementation phase that focuses on society. Therefore legislation resembles data circulation.

Legislation support technologies make this legislation process rapid and smooth. Each circulation phase has its corresponding technologies that support the phase. For instance, the search phase has document retrieval systems (e.g., [28–30]), the analysis phase has legal reasoning systems (e.g., [31, 32]), the drafting phase has statute generation systems (e.g., [33, 34]), and the enactment phase has statute databases such as e-Gov Statute Search.

The two studies presented in this chapter, coordination analysis and legal term correction, also contribute to this environment. First, coordination analysis is an important auxiliary process of syntax parsing that supports circulation. The syntax structure of Japanese statutory sentences is useful information for any practical method for the search, analysis, drafting, and enactment of statutory sentences because that information interprets them. Therefore, coordination analysis supports the data circulation of statutes through parsing. The result of coordination analysis itself also helps people to understand statutory sentences and directly accelerates the analysis of the data circulation of statutes. The proposed coordination analysis method holds a certain position among coordination analysis methods for Japanese statutory sentences because it is compatible with the hierarchical coordinate structures specific in Japanese statutory sentences and provides better predictions by utilizing LSTM-based language models.

Second, legal term correction directly supports the drafting part of the circulation of statutes. It tells legislation officers incorrectly using legal terms, making the drafting process quicker and more efficient than manually inspecting such legal terms. Additionally, legal term correction can indirectly contribute to the analysis of the circulation of statutes, which originates from the self-growth of legislation officers. Although a legal term correction system offers correction ideas, the final decision of the legal term use is left to legislation officers who must eventually consider the validity of legal term use by themselves, which is a good resource for legislation training. Once they improve their legislation skills, interpreting statutes will become more efficient. This is the contribution of legal term correction for the analysis part of the circulation of statutes.

Data Circulation of Statutes as Electronic Data

Coordination analysis forms another data circulation centralized by annotated electronic statute data (Fig. 4). Nowadays, we easily acquire statutes as electronic data from such law database systems as e-Gov Statute Search in a computer-friendly form like XML. Utilizing such data sources satisfies the data acquisition role. We then apply coordination analysis to the acquired data to obtain the coordinate structure information of the statutory sentences, which is the analysis phase. We then append the coordinate structure information to the electronic statute data and upload them to a database. The annotated statute data can be utilized for further tasks, including parsing, visualization, and simplification. That is, they have become resources for legislation support technologies. This is the implementation phase of the data circulation. Since ordinary users do not need raw coordination information in their work, it is reasonable to establish another database system for the annotated statute data in the implementation phase.

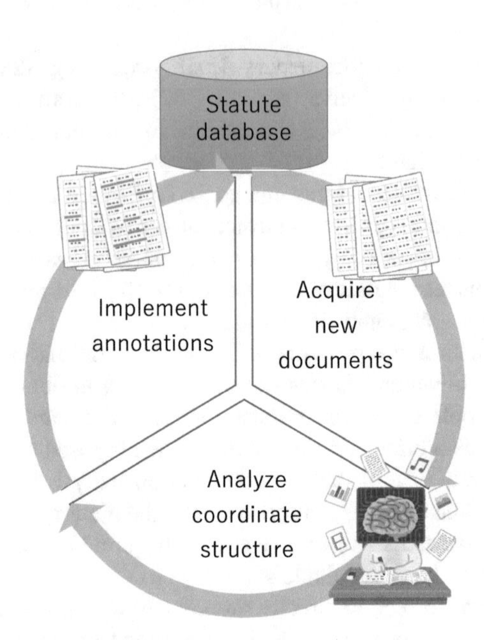

Fig. 4 Data circulation in statutes as electronic data

Data Circulation of Legal Term Correction
If we utilize the feedback of legal term correction from legislators, we can establish a data circulation centralized by correction history data. Figure 5 shows such a circulation. A legal term correction system receives a statute draft and outputs corrected ideas using the correction model, for example, the classifiers in the proposed method. In this circulation, we ask legislators to judge the correction results of the model. We put their feedback into the correction history of the system. The correction history then contributes two ways of data circulation. The first way is regarding the system, which updates the correction model from the history. The second is regarding legislation practices, where we analyze the history from the perspective of how legislators made mistakes. This analysis result is compiled as legislation tips, which teach legislators better ways to use legal terms. Both ways produce a better legislation environment so that the system can provide an improved prediction model and reduce mistakes in legislation.

LegalAI Project and its Data Circulation
For a practical activity regarding real-world data circulation, I am participating in the LegalAI project where I have been developing an online contract review system. Figure 6 shows an overview of the LegalAI system, which has three modules: anonymization, comment generation, and database.

The legal term correction methodology proposed in this study is implemented in this system as a comment generation module. Here the legal term correction model is trained by contracts. In addition to legal term correction, the system offers sev-

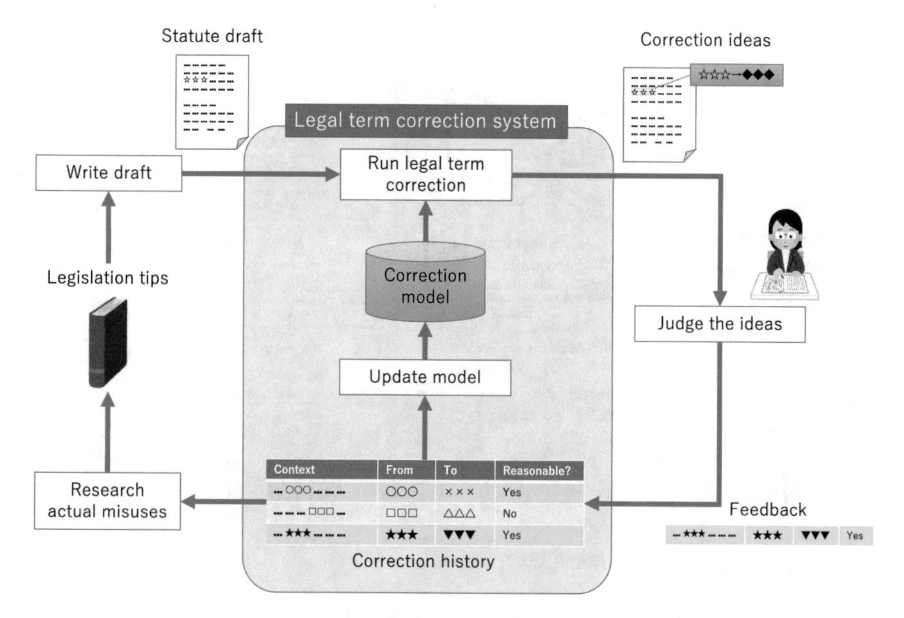

Fig. 5 Data circulation in legal term correction

eral other comment generation modules based on natural language processing technologies, such as risky clause detection, contract density judgment, and misspelling detection.

With this system, we aim to establish an organic data circulation centralized by feedback from users by automatically updating the judgment rules, the legal term set, and the classifier based on the feedback. With this plan, the system can achieve data circulation, where it acquires feedback, analyzes it to establish more sophisticated knowledge, and implements the knowledge as system updates. When we establish a data circulation that uses texts in contracts, the anonymization module is crucial. Unlike statutes, contracts often contain sensitive expressions, such as names, addresses, account numbers, transaction amounts, and so forth. Therefore, we must replace such sensitive expressions with anonymized expressions when we reuse the text data even if we are using them for training machine learning models. Even though LegalAI currently offers an anonymization module as a Web service, we do not consider it to be optimal. To maintain maximum security, we want to separate the module as an offline service that users can execute without requiring internet access.

As another utilization of real-world data regarding contracts, we aim to offer statistical facts of uploaded contracts such as frequently commented clauses, frequently appearing clauses in certain contract categories, comparisons of clause amounts between users and averages, and so on. This plan will encourage users to reflect and improve their practices when they draft contracts. Here lies another real-world data circulation for users' drafting skills. We collect contracts and their clauses in

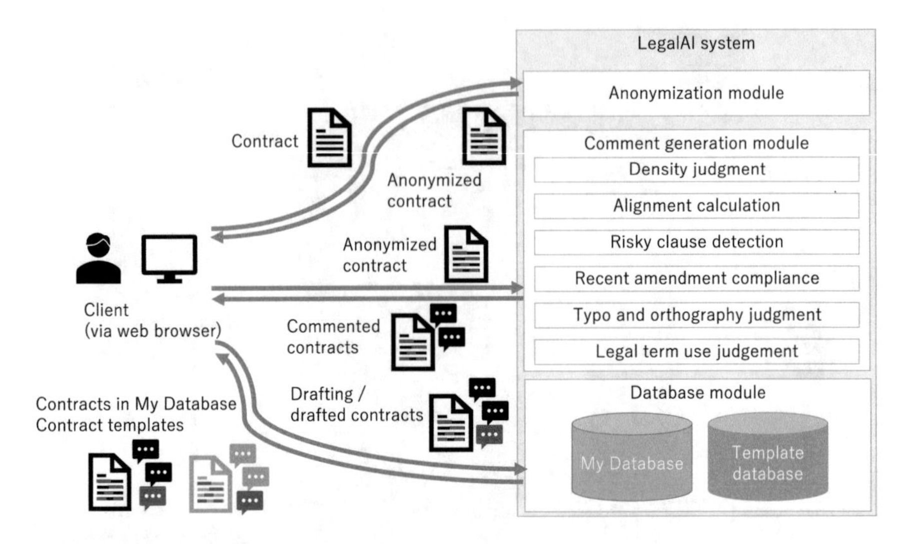

Fig. 6 LegalAI system (Cited from [35], translated)

the acquisition part and analyze statistical facts from the data in the analysis part. We then highlight the habits of users based on the statistics in the implementation part. They will eventually learn better ways to draft contracts, resulting in improved contracts.

4 Conclusion

In this chapter, we discussed two machine-learning-aided methodologies that support handling Japanese statutory sentences: coordination analysis for interpreting support and legal term correction for drafting support. Background, approach, and experimental results are briefly introduced. Next, we discussed the real-world data circulation found in legislation. Here we argued that legislation itself can be regarded as a real-world data circulation of statutes. Also, we showed that both coordination analysis and legal term correction support this data circulation by accelerating interpretation and drafting of statutes, respectively. Furthermore, each technology constructs another data circulation centered on its analysis results.

Acknowledgements The text and the table in Sect. 2.1, and Fig. 2 in Sect. 1 are based on the following paper: Takahiro Yamakoshi, Tomohiro Ohno, Yasuhiro Ogawa, Makoto Nakamura, and Katsuhiko Toyama. (2018) Hierarchical coordinate structure analysis for Japanese statutory sentences using neural language models. Vol. 25, No. 4, pp. 393–420. (C) The Association for Natural Language Processing, (Licensed under CC BY 4.0 https://creativecommons.org/licenses/by/4.0/). The text and table in Sect. 2.2 (excluding the text in Sect. 2.2.3), and Fig. 1 in Sect. 1 are based on the following paper: Takahiro Yamakoshi, Takahiro Komamizu, Yasuhiro Ogawa, and Katsuhiko

Toyama. Japanese mistakable legal term correction using infrequency-aware BERT classifier. Transactions of the Japanese Society for Artificial Intelligence, Vol. 35, No. 4, pp. E-K25 1–17, 2020. The text in Sect. 2.2.3 is based on the following paper: Takahiro Yamakoshi, Yasuhiro Ogawa, Takahiro Komamizu, and Katsuhiko Toyama. Japanese legal term correction using random forest. Transactions of the Japanese Society for Artificial Intelligence, Vol. 35, No. 1, pp. H-J53 1–14, 2020 (In Japanese).

References

1. Hoseishitsumu-Kenkyukai (2018) Workbook Hoseishitumu (newly revised), 2nd edn. Gyosei. (in Japanese)
2. Ishige M (2012) Jichirippojitsumu no tame no hoseishitsumu shokai, 2nd edn. Gyosei. (in Japanese)
3. Ficler J, Goldberg Y (2016) A neural network for coordination boundary prediction. In: Proceedings of the 2016 conference on empirical methods in natural language processing, pp 23–32
4. Hara K, Shimbo M, Okuma H, Matsumoto Y (2009) Coordinate structure analysis with global structural constraints and alignment-based local features. In: Proceedings of the 47th annual meeting of the association for computational linguistics and the 4th international joint conference on natural language processing, pp 967–975
5. Kawahara D, Kurohashi S (2006) A fully-lexicalized probabilistic model for Japanese syntactic and case structure analysis. In: Proceedings of the human language technology conference of the North American chapter of the association for computational linguistics, pp 176–183
6. Kawahara D, Kurohashi S (2008) Coordination disambiguation without any similarities. In: Proceedings of the 22nd international conference on computational linguistics, pp 425–432
7. Kurohashi S, Nagao M (1994) A syntactic analysis method of long Japanese sentences based on the detection of conjunctive structures. Comput Linguist 20(4):507–534
8. Teranishi H, Shindo H, Matsumoto Y (2017) Coordination boundary identification with similarity and replaceability. In: Proceedings of the 8th international joint conference on natural language processing, pp 264–272
9. Matsuyama H, Shirai K, Shimazu A (2012) Horei bunsho wo taisho ni shita heiretsukozo kaiseki. In: Proceedings of the 18th annual meeting of the association for natural language processing, pp 975–978. (in Japanese)
10. Bengio Y, Ducharme R, Vincent P, Jauvin C (2003) A neural probabilistic language model. J Mach Learn Res 3:1137–1155
11. Hochreiter S, Schmidhuber J (1997) Long short-term memory. Neural Comput 9(8):1735–1780
12. Kim JD, Ohta T, Tateisi Y, Tsujii J (2003) GENIA corpus—semantically annotated corpus for bio-textmining. Bioinformatics 19(suppl_1):i180–i182. https://academic.oup.com/bioinformatics/article-pdf/19/suppl_1/i180/614820/btg1023.pdf
13. Firth JR (1957) A synopsis of linguistic theory, 1930–1955. Stud Linguist Anal 1952(59):1–32
14. Ogawa Y, Yamada M, Kato R, Toyama K (2011) Design and compilation of syntactically tagged corpus of Japanese statutory sentences. In: New frontiers in artificial intelligence: JSAI 2010 conference and workshops, revised selected papers. Lecture notes in computer science, Springer, vol 6797, pp 141–152
15. Sundermeyer M, Schulüter R, Ney H (2012) LSTM neural networks for language modeling. In: Proceedings of the thirteenth annual conference of the international speech communication association, pp 194–197
16. Kudo T (2005) MeCab: yet another part-of-speech and morphological analyzer. https://ci.nii.ac.jp/naid/10019716933/
17. Cheng Y, Nagase T (2012) An example-based Japanese proofreading system for offshore development. In: Proceedings of the 24th international conference on computational linguistics, pp 67–76

18. Hitomi Y, Tamori H, Okazaki N, Inui K (2017) Proofread sentence generation as multi-task learning with editing operation prediction. In: Proceedings of the 8th international joint conference on natural language processing, pp 436–441
19. Shaptala J, Didenko B (2019) Multi-headed architecture based on BERT for grammatical errors correction. In: Proceedings of the fourteenth workshop on innovative use of NLP for building educational applications, pp 246–251
20. Takeda K, Fujisaki T, Suzuki E (1986) CRITAC—a Japanese text proofreading system. In: Proceedings of the 11th international conference on computational linguistics, pp 412–417
21. Breiman L (2001) Random forests. Mach Learn 45:5–32
22. Mikolov T, Chen K, Corrado G, Dean J (2013) Efficient estimation of word representations in vector space. In: Proceedings of international conference on learning representations, p 12
23. Mnih A, Kavukcuoglu K (2013) Learning word embeddings efficiently with noise-contrastive estimation. In: Advances in neural information processing systems, vol 26, pp 2265–2273. https://proceedings.neurips.cc/paper/2013/file/db2b4182156b2f1f817860ac9f409ad7-Paper.pdf
24. Mori K, Miwa M, Sasaki Y (2015) Sentence completion by neural language models using word order and co-occurrences. In: Proceedings of the 21st annual meeting of the association for natural language processing, pp 760–763. (in Japanese)
25. Devlin J, Chang MW, Lee K, Toutanova K (2019) BERT: pre-training of deep bidirectional transformers for language understanding. In: Proceedings of the 2019 conference of the North American chapter of the association for computational linguistics: human language technologies, pp 4171–4186
26. Kim Y (2014) Convolutional neural networks for sentence classification. In: Proceedings of the 2014 conference on empirical methods in natural language processing, pp 1746–1751
27. Tanaka M (2020) "Deta no jidai" to "puraibashi no jidai" no ryoritsu. Tech. rep, NLI Research Institute
28. Fujioka K, Komamizu T, Ogawa Y, Toyama K (2019) Bubunkozo wo mochiita ruijireiki no kensaku. In: Proceedings of DEIM 2019, p 4 (in Japanese)
29. Sugathadasa K, Ayesha B, de Silva N, Perera AS, Jayawardana V, Lakmal D, Perera M (2019) Legal document retrieval using document vector embeddings and deep learning. In: Intelligent computing, pp 160–175
30. Yoshikawa K (2017) Identifying relevant civil law articles with judicial precedents. In: Proceedings of the eleventh international workshop on juris-informatics, pp 142–155
31. Batsakis S, Baryannis G, Governatori G, Tachmazidis I, Antoniou G (2018) Legal representation and reasoning in practice: a critical comparison. In: the thirty-first annual conference on legal knowledge and information systems (JURIX 2018). IOS Press, Amsterdam, pp 31–40
32. Chhatwal R, Gronvall P, Huber-Fliflet N, Keeling R, Zhang J, Zhao H (2018) Explainable text classification in legal document review a case study of explainable predictive coding. In: 2018 IEEE international conference on big data (Big Data), pp 1905–1911
33. Debaene S, van Kuyck R, Buggenhout BV (1999) Legislative technique as basis of a legislative drafting system. In: The twelfth conference of legal knowledge based systems (JURIX 1999), pp 23–34
34. Hafner CD, Lauritsen M (2007) Extending the power of automated legal drafting technology. In: The twentieth annual conference on legal knowledge and information systems (JURIX 2007), pp 59–68
35. Yamada H (2020) Potentiality of AI technology for practical issues in contract drafting and review. In: Proceedings of the 34th annual conference of the Japanese society for artificial intelligence, p 4. (in Japanese)

Research of ICT Utilization for the Consideration of Townscapes

Mari Endo

Abstract The townscape reflect the social background and values of people of each period. Although many buildings and towns were destroyed, people have to know those values. On the other hand, a huge number of photos had been shot since cameras became popular. This chapter deals with the values of those film photos as records of townscapes. In this research, two applications were developed to recognize the change of townscapes by using old photos and carried out demonstration experiments. Results of experiments are shown. In addition, the change of values of photos by implementation as contents of applications and the possibility of photo archives are considered.

1 Introduction

As time changes, townscapes change, since townscapes reflect lifestyles, thoughts, and values of societies of each period. In Japan's period of high economic growth, in urban areas, large-scale housing complexes were constructed and multi-story apartments and new houses were built following the increase of population and household numbers. Therefore, townscapes changed drastically, and accordingly, many historical buildings and remains were destroyed. The Japanese government established the Cultural Property Protection Law in 1950 and traditional building conservation areas in 1975. They put effort into the preservation and inheritance of precious cultural townscapes. On the other hand, there are precious values like cultural history of local areas and life histories of people living in buildings and townscapes that are not designated cultural properties. We should evaluate not only the remarkable value of cultural property, but also the value from the viewpoint of regional perspectives. To do so, we have to improve literacy to evaluate the value of cultural properties we look at. We call this literacy, the "visual literacy" of a townscape.

The aim of this research is to make people focus their attention on the historical significance of a townscape to consider the value from the regional perspective. We

M. Endo (✉)
Department of Global and Media Studies, Kinjo Gakuin University, Nagoya, Japan
e-mail: marie@kinjo-u.ac.jp

K. Takeda et al. (eds.), *Frontiers of Digital Transformation*,
https://doi.org/10.1007/978-981-15-1358-9_13

provide a means so that people can find out and consider the value of the townscape by themselves through old photos and information technology. Applications are developed to compare the current townscape with the old townscape by using information communication technology including augmented reality technology. These applications could provide people with the chance to become conscious of the value of the townscape because people can experience the change more instinctively in a real town than just the knowledge from books, museums, and so on.

2 Visual Literacy of Townscape

Visual literacy [1, 3] was defined by Debes[1] as follows.[2]

> Visual literacy refers to a group of vision competencies a human being can develop by seeing and at the same time having and integrating other sensory experiences. The development of these competencies is fundamental to normal human learning. When developed, they enable a visually literate person to discriminate and interpret the visible actions, objects, and symbols natural or man-made, that he encounters in his environment. Through the creative use of these competencies, he is able to communicate with others. Through the appreciative use of these competencies, he is able to comprehend and enjoy the masterworks of visual communication.

As an intelligible example of visual literacy, there is the William's[3] case that she introduced in her book[4] [2] as follows.

> Many years ago I received a tree identification book for Christmas. I was at my parents' home, and after all the gifts had been opened I decided I would identify the trees in the neighborhood. Before going out, I read through some of the identification clues and noticed that the first tree in the book was the Joshua tree because it only took two clues to identify it. … I would know if I saw that tree, and I've never seen one before. … I took a walk around the block … at least 80 percent of the homes had Joshua trees in the front yards. And I had never seen one before! Once I was conscious of the tree – once I could name it – I saw it everywhere.

To come in sight is different from to recognize. People become able to be aware of things after they recognize them by acquiring knowledge. This can be implied from Williams' case.

Debes mentioned in his definition of visual literacy that a human being can develop it by seeing and at the same time having and integrating other sensory experiences. Seeing is simply seeing trees and other sensory experience is reading a book about the Joshua tree in Williams' case. Williams could develop her visual literacy about the Joshua tree by integrating seeing and other sensory experiences (Fig. 1).

In the case of the visual literacy of a townscape, people with it can find and recognize the historical significance of a townscape when they see similar buildings

---

[1]Fransecky and Debes [1], USA, writer, educator, founder of IVLA.

[2]Fransecky and Debes [1], *Visual Literacy: A Way to Learn-A Way to Teach*, AECT Publications.

[3]Williams [2], USA, writer, educator, graphic designer.

[4]Williams [2], The Non-Designer's Design Book, *Peachpit Press.*

A human being can develop visual literacy by seeing and at the same time having and integrating other sensory experiences. (Debes)

Fig. 1 Development of the visual literacy in Williams' case

Visual literacy of Townscape:

The ability for finding and recognizing the historical significance of the townscape

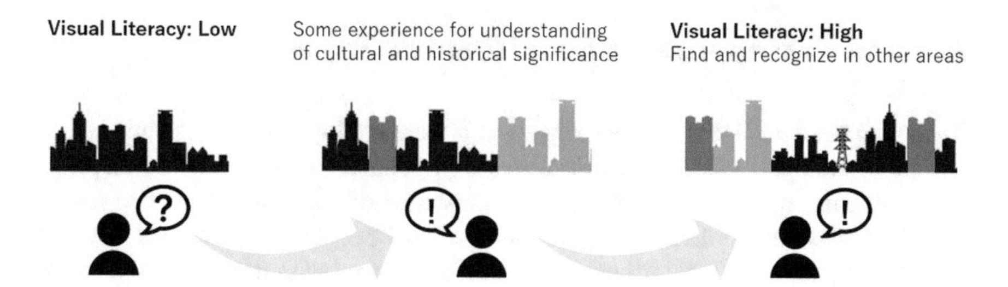

Fig. 2 Development of the visual literacy of a townscape

and townscapes after they recognize the historical significance of the townscape on a certain building or area. In this research, the ability for finding and recognizing the historical significance of a townscape is defined as the visual literacy of a townscape (Fig. 2).

3 Applications for Developing the Visual Literacy of a Townscape

In this research, we focus on old photos of a townscape as a way for developing the visual literacy of the townscape. Two kinds of smartphone applications by using old photos were developed. Both applications lead people to places where old photos were taken and enable people to compare the old townscape with the current real townscape on a smartphone. Sensory experiences other than those that Debes mentioned are provided through these applications. People are expected to be

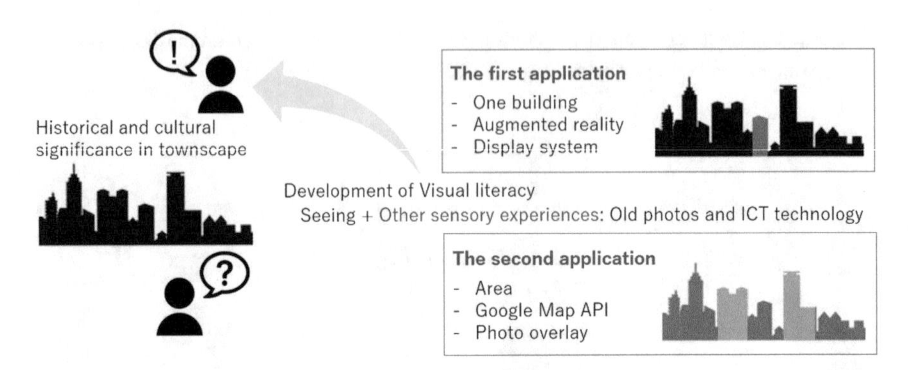

Fig. 3 Development of the visual literacy of a townscape through applications

able to develop their visual literacy of a townscape through these applications and become able to recognize the historical significance of the townscape (Fig. 3).

3.1 The First Application for a Historical Site in a Townscape

We tend to see a townscape innocently. Even if there is a historical site, many people pass by it without considering its historical significance. The first step to recognize the historical significance is to know the existence of such sites in a townscape. Therefore, the first application's function is to lead people to a historical site and display old photos on a smartphone. In addition, a display system connected with the application is developed to attract users' interests (Fig. 4).

Leading Users to the Place Where Old Photos Were

Users are lead to a location where old photos were actually taken by displaying the direction and the distance to the sites using augmented reality technology on a smartphone. When the users' locations are far from the site, the application indicates the distance and the direction to it. Users will approach the site following the directions on the display, and walk around (Fig. 5).

Displaying Old Photos on a Smartphone

When the distance and direction are met, the application shows the old photos. Users can learn about the history of the site and compare the old photos with the current situation (Fig. 6). When users tap the information button, texts explaining the history of the site are displayed.

Planning for Users' Participation with the Display System

Not only displaying the old photos, but also adding the arrangement of users' participation attracts the users' interests to the site. Thus, a function for taking pictures and uploading them to a server is implemented in the display system. The EXIF data of

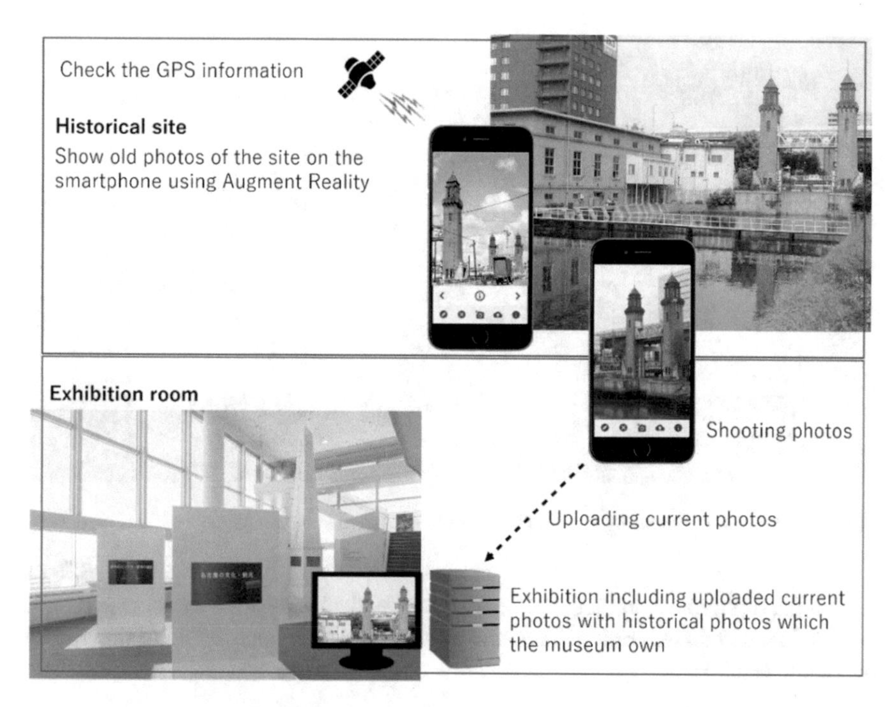

Fig. 4 System image of the application and the display system

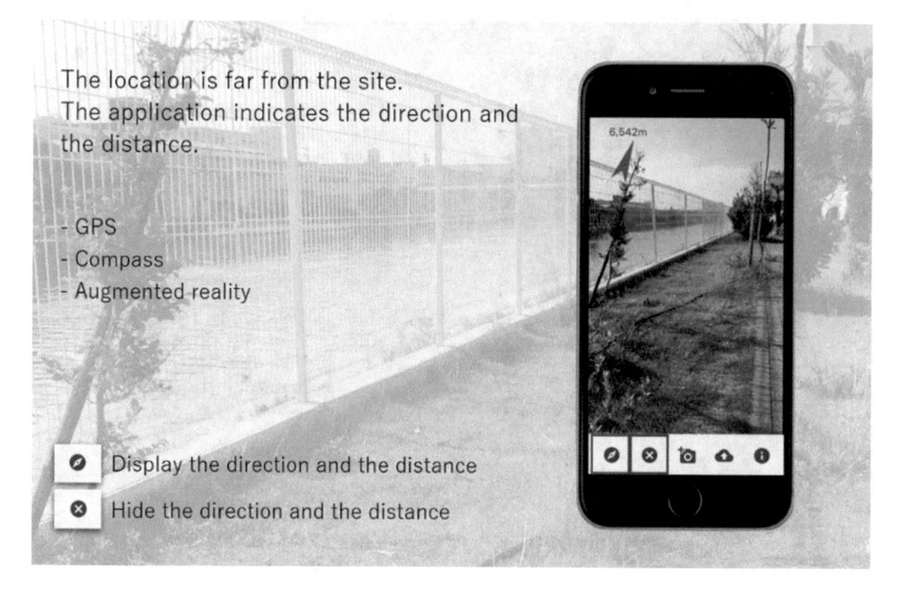

Fig. 5 Interface for leading users

Fig. 6 Interface for comparing the old photos and the current situation

an uploaded photo are immediately added to the display system, then those photos are displayed at an exhibition. This makes users feel they contributed to the history of the site (Fig. 7).

Fig. 7 Display system

3.2 *Experiment and Result of the First Application*

A demonstration experiment using the first application was conducted in November 2016. Feedbacks were collected in the form of a questionnaire which was filled in by 63 people around the site. At the same time, both old photos and user-contributed photos were exhibited and feedbacks were collected in the form of a questionnaire which was filled in by 26 people at a gallery. In the results of this questionnaire-based survey, over 80% of the users of the application answered they could learn more about the site. The reasons why they wanted to visit were that they were interested in the site and the history of the area. We can see that this application attracted not only the temporal interests of users in the application, but also certainly the historical significance of the site and the area. They also answered that they found the attractiveness of the old photos themselves. This should hopefully lead to the preservation of old photos. In addition, over 80% of the users answered they wanted to visit the exhibition gallery and to see the photos contributed by themselves. This could also be the reason why users' interests were attracted to the application.

3.3 *The Second Application for Local Area History*

The experiment of the first application clarified the possibility of expanding the application to other areas using old photos. So, the second application was developed. While the first application was for one specific site, the second one expands the target to a local area. Users are expected to be able to recognize the change of a townscape and the historical significance by using this application at various locations throughout the town. The main points of modification from the first application are as follows.

Displaying Location Information by Google Map API
An augmented reality technology is used to lead users to the sites. On the other hand, the target of this application is the whole area. It should be more effective if users walked around voluntarily and found locations where old photos were taken than the application leading them. Therefore, location information of users and that where photos were taken are displayed by using the Google Map API (Fig. 8).

Addition of Photo Overlay Function
The second application is equipped with a function to overlay an old photo on the camera image. In addition, users can change the degree of transparency of the overlaid photo. This function is added so that users can find locations where old photos were taken, compare old photos with the real townscape in more detail, and recognize the change of townscape and the historical significance (Fig. 9).

Fig. 8 Interface for displaying location information

Fig. 9 Photo overlay function

3.4 *Experiment and Result of the Second Application*

A demonstration experiment was conducted using the second application in May 2019. Research subjects of the experiment were 8 university students and 6 volunteer guides of a local area. The application was presented to them and feedbacks were collected in the form of a questionnaire. Volunteer guides answered that there is a possibility that the way of their guidance could be expanded with this application. On the other hand, there were some answers that they felt uneasy about the controllability of the application and the explanation to visitors. Meanwhile, some university students answered that they noticed it was interesting to recognize the difference between the current townscape and the old townscape. This indicates that the application attracts users' interests in recognizing the change of a townscape.

4 Records by Photos in the Circulation World

People find various values in things and phenomena in the real world, and they circulate and reflect on the real world. Townscapes reflect the values of people and the social background of each period, too. Values which people find in townscapes are artistic values of buildings and historical values of the area. They are protected as cultural properties because people recognize those values. On the other hand, there are many buildings and areas whose values are not recognized, because they are not researched nor recorded. As mentioned in the previous sections, many historical buildings and remains are destroyed because of the lack of our recognition of their historical and cultural values. The applications introduced in Sect. 3 were intended to lead people to understand them. The influence on the real world by this research is to lead people to become aware of new knowledge and the value of visual literacy of townscapes. In addition, this application adds the value as recorded data of townscapes (Fig. 10). This is helpful for not only conserving old photos from the past, but also storing current photos. In this section, we focus on adding value to photos by data circulation in the real world.

4.1 *Photos as Records of Townscape*

Camera became widespread in the second half of the nineteenth century, and since then many photos were taken around the world. The number of photos taken by film cameras was about 8.5 billion throughout the year in the year 2000 [4]. This means that about 2,500 photos were taken per second. However, the number of film photos decreased rapidly because of the spread of digital cameras. The days when people shared photos with families and friends after shooting and development have gone away. Furthermore, the characteristics of photos changed to sharing digital data with

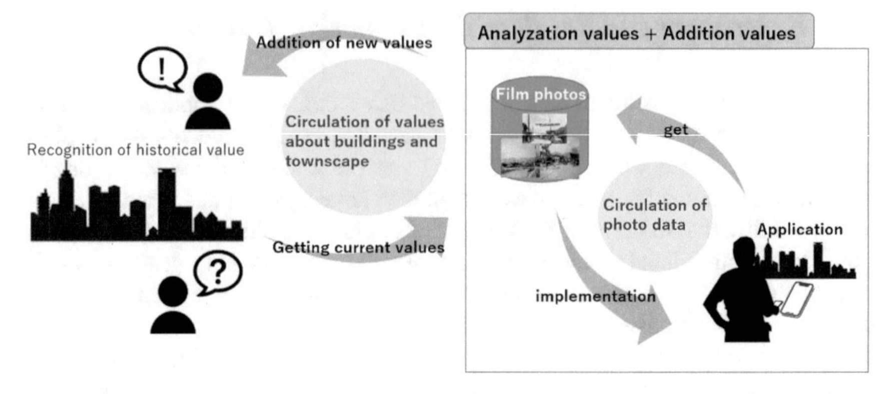

Fig. 10 Addition of values by circulations

people all over the world through social networking services, like Facebook, Twitter, and Instagram. Although most of the film photos taken before the year 2000 are stored in each household, they are in the process of disposing because of their degradation or change of generations. These photos include snapshots that were taken in touristic areas and in daily life. In addition, they could be precious records for understanding the lifestyle and the townscape in those days, because there are many photos that have shot the change of townscapes.

John Urry,[5] who was a sociologist, said in his book[6] [5] as follows:

> What is sought for in a holiday is a set of photography images, which have already been seen in tour company brochures or on TV programmes,… It ends up with travellers demonstrating that they really have been there by showing their version of the images that they had seen before they set off.

As Urry said, many tourists go to see touristic sites which are introduced in brochures or on TV programs, then they take photos from the same angle as introduced in them. This action of tourists implies that there are many photos shot at the same location and from the same angle at different timings. These photos could be considered like time-lapse photos taken from the past to the present (Fig. 11).

As mentioned in previous sections, the developed applications allow us to compare the old photos with the current townscape at the same location and from the same angle. This plays the same role as the brochures and TV programs in Urry's example.

4.2 Examples of Utilization of Recorded Photos

There are many cases that recorded photos of the change of townscapes served as references for restoring precious cultural properties. For example, many photos of

[5]Urry [5], (1946–2016), UK, Sociologist.

[6]Urry [5], *The Tourist Gaze*, SAGE Publications Ltd.

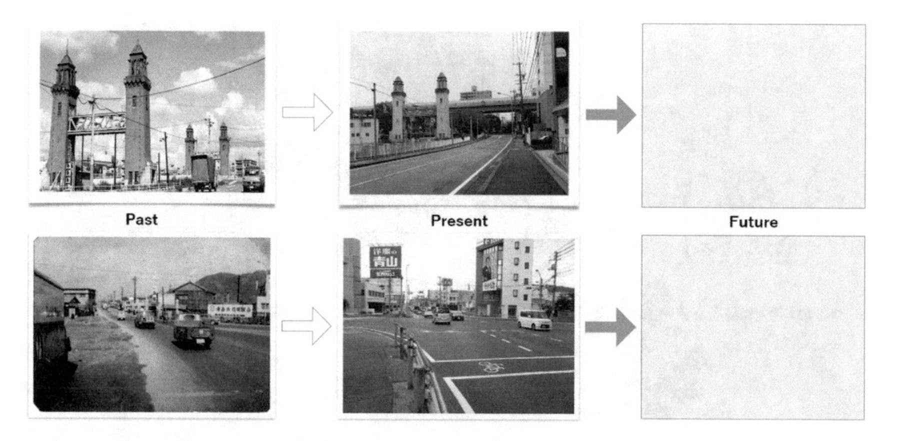

Fig. 11 Photos from the past to the present and the future

Nagoya Castle from the latter part of the Edo period have been preserved. Although the castle was destroyed by fire during World War II, it was restored referring to old photos shot in the early Showa period [6, 7]. In recent years, Kawakami[7] and her team has been engaged in the digital restoring project of Shuri-jo Castle that was destroyed by fire in 2019 [8]. They called to people for providing photos and videos of the castle, and have restored the three-dimensional shapes from photos by using Structure from Motion technology on a computer. These examples showcase the value as data for restoration was added to townscape photos. Similarly, the three-dimensional shapes of old buildings and townscapes could be restored by making use of computer vision and AI technologies and applied for augmented reality or virtual reality environments. These technologies will have a big impact not only on entertainment, but also on education or sociology. On the other hand, more recorded photos are needed for restoring more precisely. The application developed for visual literacy in this research are expected to be helpful for collecting recorded references.

4.3 *Adding Values to Photos by Applications*

Values of photos changed through the developed applications in this research as follows: (Fig. 12).

a. To digital data from analog film photos
   Analog film photos archived in public corporations and private households were digitized.
b. To contents by being used in the applications
   When and where a photo was shot were checked, then annotated as contents for the applications. The location where a photo was shot was confirmed manually.

[7] *Rei Kawakami, JPN, Research Associate Professor, Tokyo Institute of Technology.*

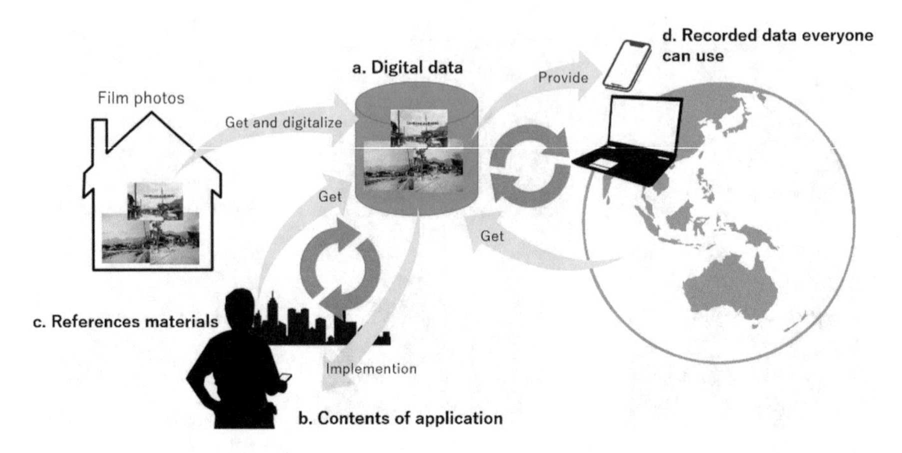

Fig. 12 Circulation of values of photos

In the future, the development of AI technology may allow us to specify the location by analyzing a lot of photos.

c. To references for comparing old photos with the current townscape
Users could find the difference between the old townscape and the current townscape by comparing them on the application.

d. To recorded data everyone can use
Newly recorded photos are shot at a similar location from a similar angle with an old photo, since users tend to shoot photos referring to old photos in the application. In addition, new digital photos are accumulated in the display server by the first application introduced at Sect. 3.1, then displayed at an exhibition. They could become recorded data that everyone can use by archiving and providing freely in the future.

As noted above, the values of film photos have changed through the development and the implementation of the proposed applications. Furthermore, new values will be created by circulating back to the real world. This circulation should be helpful for not only recognizing the cultural and historical significance, but also other research and development like the restoring project mentioned in Sect. 4.2, local tourism, and so on.

5 Conclusions

In this chapter, the value that photos have and created by data circulation in the real world have been discussed. Most photos were taken in the real world record only private scenes and memories. However, some of them have the value of recorded data of cultural and historical significance. By collecting and utilizing them, it could be applied to new technologies, like restoring buildings or constructing three-dimensional shapes for lost buildings. This research showed that a new value as recorded data is added to film photos by developing applications using old film photos. In addition, there is a possibility that more recorded photo data could be collected and new values could be created with the use of the proposed applications.

On the other hand, we have to pay attention to copyright and privacy issues when we use those photo data. It is important to consider more accessible archiving and to thoroughly explain the merit of the utilization of recorded photo data to the providers.

References

1. Fransecky RB, Debes JL (1972) Visual literacy: a way to learn-a way to teach. AECT Pulbications
2. Williams R (2014) The non-designer's design book. Peachpit Press
3. Motoyama K, Endo J, Sadakuni S, Suzuki N, Mizuuchi T (2018) Visual literacy studies. Chunichi-bunka Co., Ltd (in Japanese)
4. Good J (2019) How many photos have ever been taken?. 1000 memories blog (2019.3.24 Read) https://web.archive.org/web/20130116130505/http://blog.1000memories.com/94-number-of-photos-ever-taken-digital-and-analog-in-shoebox
5. Urry J (2011) The tourist gaze. SAGE Publications Ltd
6. Special historic site Nagoya Castle, "Nagoya Castle Keep Reconstruction Project" (in Japanese) (2019.11.11 read) https://www.nagoyajo.city.nagoya.jp/learn/tenshu/
7. Independent Administrative Institution National Research Institute for Cultural Properties, "Nagoya jo honmaru gotten" (in Japanese) (2019.11.11 read) https://www.tobunken.go.jp/image-gallery/nagoya/index.html
8. Team OUR Shurijo, "Shuri Castle Digital Reconstruction" (in Japanese) (2019.11.11 read) https://www.our-shurijo.org/

Measuring Efficiency and Productivity of Japanese Manufacturing Industry Considering Spatial Interdependence of Production Activities

Takahiro Tsukamoto

Abstract We developed anew spatial autoregressive stochastic frontier model for the panel data.The main feature of this frontier model is aspatial lag term of the explained variables, ajoint structure of aproduction possibility frontier with amodel of technical inefficiency.This model addresses the spatial interdependence, the heteroskedastic technical inefficiency.This study applies the maximum likelihood methods by considering the endogenous spatial lag term.The proposed model nests many existing models. Further, an empirical analysis using data on the Japanese manufacturing industry is conducted, the existing models are tested against the proposed model, which is found to be statistically supported.The findings suggest that estimates in the existing spatial, non-spatial models may exhibit bias because of lack of determinants of technical inefficiency, as well as aspatial lag.This bias also affects the technical efficiency score, its ranking.Our model can analyze industrial policies including industrial cluster policies, also it is expected that, by devising the spatial weight matrix, it will be possible to analyze what kind of industrial agglomeration form is desirable for improving productivity.

1 Economics and Concept of Real-World Data Circulation

In order to create a cycle of the process of creating social values, it is important to consider the three aspects of data "acquisition," "analysis," and "implementation." "Acquisition" refers to obtaining digital data through an observation of various real-world phenomena. "Analysis" refers to analyzing real-world data using information technology, statistical techniques, and so on. "Implementation" refers to the social

---

This chapter is based on [1], with additions and corrections.

T. Tsukamoto (✉)
Graduate School of Economics, Nagoya University, Furo-cho, Chikusa-ku, Nagoya, Aichi 464-8601, Japan
e-mail: t.tsukamoto@nagoya-u.jp

K. Takeda et al. (eds.), *Frontiers of Digital Transformation*,
https://doi.org/10.1007/978-981-15-1358-9_14

contribution based on the results of the analysis. The concept of creating social values by circulating these three phases is at the heart of Real-World Data Circulation (RWDC).

Although economics has several definitions, it generally refers to the study of any phenomenon related to goods, people, or money. Economics seeks to elucidate the structure of our society and develop it. It tries to maximize people's utility by allocating limited resources efficiently. Therefore, it is not an exaggeration to say that the purpose of economics is to create social values. In economics, it is true that many studies focus on the "analysis" phase, but studies on "acquisition" and "implementation" have also been conducted. A typical example of research on "acquisition" is the system of national accounts. The gross domestic product (GDP), a well-known indicator of economic activity, is one of the items of the system of national accounts. Hereafter, it is expected that a more detailed and accurate information will be collected in a timely manner by utilizing information technology. In economics, "implementation" often means policy implementation in the real world. Several studies also focus on policy implementation. In the field of public economics, frameworks and policies that achieve the economically desirable state (Pareto optimum) in the presence of market failure have long been analyzed. Several recent studies have also been conducted on feasibility in the real world. The carbon emissions trading and the congestion taxes are good examples of the real-world implementations of the results of economics research.

Even if we focus on the "analysis" phase, we should still imagine the creation of the potential three processes of "acquisition," "analysis," and "implementation." Since economics is an academic field related to the human society, it embraces both the scientific and practical aspects. While research on the development of economics as science is important, it is also our mission to orient our research toward societal development. In order to build a better society, it is very valuable to think about the cycle of the three processes of "acquisition," "analysis," and "implementation," instead of only the scientific aspect.

The topics covered in this chapter are productivity, efficiency, and production technology, all of which are extremely important to our society. Improving productivity through improved efficiency or technological progress implies that more output can be obtained with a certain input. This is desirable in terms of the effective use of resources, and it is expected to increase the real wages of workers. These are the typical potential social values of our work.

2 Background and Aims

The stochastic frontier models are often used to estimate the production functions at the establishment-, plant-, firm-, or regional levels. These models were first proposed by Aigner et al. [2] and Meeusen and van den Broeck [3] almost at the same time. The feature of this model is a composed error structure consisting of two variables: a random variable that captures noise and a random variable that explains technical inefficiency. When the production function is estimated using the ordinary least squares (OLS), the estimated production function is interpreted as the average production function. However, if we use a stochastic frontier model, then the estimated production function can be interpreted as the PPF. The stochastic frontier model can be used in empirical studies based on the economic models, given that the PPF illustrates the production function in microeconomics. However, since the production activities may be interdependent with neighboring production activities by various externalities or input–output networks such as a supply-chain network, observations on production entities are likely to correlate depending on their geographical distances. Thus, it is very doubtful that the assumption of spatial independence, as usual stochastic frontier models assume, is appropriate. Knowledge spillover is one of the typical examples of spatial externalities. Knowledge spillover promotes the imitation and innovation of a production technology. The spatial proximity between entities promote knowledge spillover. Therefore, if knowledge spillovers are present, production activity is spatially interdependent. Also, from the fact that production activities affect the labor market and the intermediate goods market in that area and its surrounding areas and vice versa, it turns out that production activities are spatially dependent. In addition, networks through Input–Output have been constructed. For example, the Toyota Group's offices and plants are located mainly in Aichi Prefecture and its surrounding prefectures, and they have established a large industrial linkage network. These network linkages facilitate an active internal exchange of human resources and technologies. Hence, these regions are considered mutually dependent.

In the field of spatial econometrics, there has been a constant evolution of econometric models to address spatial interdependence. The basic spatial econometric models are as follows: the spatial autoregressive model (SAR), which includes a spatial autocorrelation structure (spatial lag) of explained variables; the spatial error model (SEM), which includes a spatial autocorrelation structure in error terms; and a spatial lag of X model (SLX), which includes an autocorrelation structure (spatial lag) of the explanatory variables, and their integrated models [4]. SLX is a model that captures the local spillover of the explanatory variables. Since there are no notable estimation problems, SLX can be estimated by OLS. SEM and SAR capture global spillovers and cannot be estimated by OLS. As Glass et al. emphasize [5], in SAR, spillovers are explicitly adopted and are related to the independent variables. In LeSage and Pace, a spatial spillover is defined as non-zero cross-partial derivatives $\partial y_j/\partial x_i, j \neq i$ [6], which means that changes to the explanatory variables in region i impact the explained variables in region j. In this definition, SAR allows a spatial spillover, while SEM does not allow this spillover. The inability of SEM to address the spatial spillover reduces its appeal [7]. Fingleton and López-Bazo criticize SEM because it absorbs externalities into random shocks [8].

Many empirical studies have focused on the magnitude and existence of a spatial spillover [9]. If an analyst is interested in spatial spillover, SAR is preferred. However, the SAR has concerns about omitted variables with a high degree of spatial interdependence, which could lead to an overestimation of the magnitude of externalities [7].

Since the 2010s, the literature has witnessed an evolution of stochastic frontier models considering spatial interdependence (spatial stochastic frontier models). Druska and Horrace conducted the first study on the spatial stochastic frontier models [10]. The authors estimated a spatial error production frontier panel data model using the generalized methods of moments by integrating Kelejian and Prucha's stochastic frontier model and SEM [11]. They calculated the time-invariant technical inefficiency and concluded that the consideration of spatial correlation affects the technical efficiency (TE) score and its ranking. Adetutu et al. estimated the stochastic frontier model that incorporates the structure of SLX [12]. We can directly apply the estimation procedures for usual non-spatial stochastic frontier models to the stochastic frontier model with an SLX structure. Glass et al. proposed a spatial autoregressive stochastic frontier model (SARSF) for panel data [5] by integrating SAR and a half-normal stochastic frontier model proposed by Aigner et al. [2]; they estimate using the ML methods. Glass et al. introduced the concept of an efficiency spillover [5]. Their inefficacy term was homoskedastic. Ramajo and Hewings developed an SARSF model for panel data with a time-varying decay efficiency specification [13], by extending Battese and Coelli's model [14]. This SARSF model permitted the inefficiency to increase or decrease exponentially based only on a scalar measuring the yearly rate of the technological catch-up. Gude et al. developed a heteroskedastic SARSF model with a feature wherein the parameter of the spatial lag term is heteroskedastic [15]. Their inefficacy term was homoskedastic. Concerning the non-frontier SAR, an omission of spatially dependent variables in these existing stochastic frontier models with a SAR structure may lead to an overestimation of the magnitude of externalities.

In the literature of non-spatial stochastic frontier models, it has been considered important to estimate considering the determinants of technical inefficiency. Since it is possible to estimate the score of technical efficiency for each unit in stochastic frontier models, it is natural to try to find the determinants of the score. In early studies, such as Kalirajan [16] in 1981, to know the determinants of technical inefficiency, a "two-stage approach" was adopted in which technical inefficiency was first estimated using a stochastic frontier model, and, subsequently, the estimated value was regressed with factor variables. However, the second-stage contradicts the assumption in the first-stage, that is, the probability distribution of random variables explaining the technical inefficiency is mutually independent. Subsequent studies (Kumbhakar et al. [17] in 1991; Reifschneider and Stevenson [18] in 1991; Caudill and Ford [19] in 1993; Huang and Liu [20] in 1994; Caudill et al. [21] in 1995; Battese and Coelli [22] in 1995) developed a (non-spatial) "single-stage approach" that simultaneously estimates the stochastic frontier and the determinants of technical inefficiency. As Kumbhakar and Lovell mentioned, if there are determinants of technical inefficiency which correlate with explanatory variables (input quantity of production function), the parameter estimates in usual stochastic frontier models will have a bias [23]. This of course also affects the TE score. Therefore, if the technical inefficiency is not completely randomly determined or if the determinants of technical inefficiency may exist, then it would be appropriate to prefer the single-stage approach, regardless of whether analysts are interested in the determinants of technical inefficiency. Battese and Coelli's model proposed in 1995 [22] is a single-stage approach model adopted by the largest number of empirical studies, such as Fries and Taci [24], Srairi [25], and Saeed and Izzeldin [26]. Given this background, it is expected that stochastic frontier model with an SAR structure introducing a model of technical inefficiency can correctly estimate parameters including parameters related to the spatial interdependence. Pavlyuk mentioned the spatial stochastic frontier model that simultaneously estimates the determinants of technical inefficiency as "one possible spatial modification of the stochastic frontier" [27]. However, Pavlyuk presented neither the estimation method nor the estimates. We do not know any study estimating single-stage approach stochastic frontier models with the SAR structure.

We develop a spatial stochastic frontier model with the SAR term and a feature of Battese and Coelli's model [22], which simultaneously estimates the determinants of technical inefficiency. The proposed model can identify the cause of technical inefficiency. It also has the merit of dealing with the omitted-variable bias because of the lack of the determinants of technical inefficiency and the spatial lag. Moreover, the proposed model nests many existing spatial and non-spatial stochastic frontier models. The model selection can be easily done by a statistical test for the nested structure.

3 Model

Our proposed SARSF production model for panel data that incorporates a model of technical inefficiency is as follows:

$$y_{it} = \boldsymbol{x}_{it}^{'}\boldsymbol{\beta} + \rho \sum_{j=1}^{N} w_{ij}^{t} y_{jt} + v_{it} - u_{it}, i = 1, 2, \ldots, N, t = 1, 2, \ldots, T_i,$$

$$v_{it} \sim i.i.d.\mathrm{N}\left(0, \sigma_v^2\right), u_{it} \sim i.i.d.\mathrm{N}^{+}\left(\mu_{it}, \sigma_u^2\right), \mu_{it} = \boldsymbol{z}_{it}^{'}\boldsymbol{\delta}. \quad (1)$$

In the above equations, y_{it} is a scalar output of the producer i in the period t, and $\boldsymbol{x}_{it}$ is a vector of inputs used by the producer i in the period t. $\boldsymbol{z}_{it}$ is a $(m \times 1)$ vector of determinants that may generate technical inefficiency, including both producer-specific attributes and environmental factors. $\boldsymbol{\delta}$ is a $(m \times 1)$ vector of unknown parameters. The second term in the right-side-hand of Eq. (1) represents the SAR term that captures the spatial dependency. w_{ij}^{t} is the ij element of the spatial weight matrix in the period t. The whole spatial weight matrix $\boldsymbol{W}$ is a block diagonal matrix of $\left\{\boldsymbol{W}^1, \boldsymbol{W}^2, \ldots, \mathbf{W}^{T_i}\right\}$, where $\mathbf{W}^t = \{w_{ij}^t\}$, and ρ is an unknown parameter associated with the SAR term. The parameter space of ρ is $(1/\omega_{min}, 1/\omega_{max})$, where ω_{max} and ω_{min} are the smallest and largest eigenvalues of $\boldsymbol{W}$, respectively. v_{it} is random noise and u_{it} represents the technical inefficiency. It is assumed that v_{it} and u_{it} are independent. The spatial weight matrix $\boldsymbol{W}$ is a non-negative and non-stochastic matrix that describes the strength of the relationship between the cross-sectional units. To eliminate the direct influence on itself, the diagonal matrix is set to 0. The spatial econometrics literature has proposed various spatial weight matrix specifications such as a binary adjacency matrix or a matrix defined as the decreasing function of the geographical or economic distance between regions. If $\boldsymbol{W}$ is row-normalized (each row sum is set to 1), the product $\boldsymbol{Wy}$ of the spatial weight matrix and the explanatory variables can be interpreted as a weighted average of the explanatory variables. Therefore, many spatial econometric studies have used the row-normalized spatial weight matrices [28]. When elements of a spatial weight matrix are defined as inverse distances, the row-normalization eliminates the influence of the distance unit.

In the usual linear models, $\boldsymbol{\beta}$ represents the marginal effects, which can be interpreted as elasticity when the variables are logarithmic values. However, as noted by Kelejian et al. [29], LeSage and Pace [6], and Glass et al. [5], among others, $\boldsymbol{\beta}$ does not represent the marginal effects in models with an SAR structure. In our model, the partial derivatives matrix with respect to the r th explanatory variables $\boldsymbol{x}^r$ is:

$$\begin{aligned}\frac{\partial \boldsymbol{y}}{\partial \boldsymbol{x}^{r'}} &= (\boldsymbol{I}_{NT} - \rho \boldsymbol{W})^{-1}\beta_r \\ &= \left(\boldsymbol{I}_{NT} + \rho \boldsymbol{W} + \rho^2 \boldsymbol{W}^2 + \rho^3 \boldsymbol{W}^3 + \rho^4 \boldsymbol{W}^4 + \ldots\right)\beta_r \end{aligned} \tag{2}$$

where $\boldsymbol{y} = \{y_{it}\}$ is a vector of outputs, and β_r denotes the r th parameter of $\boldsymbol{\beta}$. The marginal effect varies over observations. Every diagonal element of the matrix denotes the marginal effect of the own explanatory variable, which is called a direct effect. Every off-diagonal element of the matrix denotes the marginal effect of the not-own explanatory variable, which is called an indirect effect. LeSage and Pace proposed an average of the diagonal elements of the matrix as the summary statistics of the direct effects [6]. However, since the average lacks a substantial amount of information, it is necessary to report the indices representing the dispersion of the direct effect (e.g., maximum and minimum values).

The proposed model nests many existing spatial and non-spatial stochastic frontier models. If $\rho = 0$, then the proposed model turns to be equivalent to the model suggested by Battese and Coelli [22]. If z_{it} includes only a constant term, then our model becomes a spatial stochastic frontier model assuming a truncated normal distribution as the distribution that represents technical inefficiency. If $\boldsymbol{\delta} = 0$, then our model becomes equivalent to the SARSF model assuming a homoskedastic half-normal distribution as a distribution that represents the technical inefficiency proposed by Glass et al. [5]. If $\rho = 0$ and z_{it} has only a constant term, then our model will be a non-spatial stochastic frontier model assuming a truncated normal distribution as the distribution that represents the technical inefficiency proposed by Stevenson [30]. If $\rho = 0$ and $\boldsymbol{\delta} = 0$, our model becomes equivalent to a non-spatial stochastic frontier model assuming a homoskedastic half-normal distribution as a distribution that represents the technical inefficiency proposed by Aigner et al. [2]. The spatial lag term $\sum_{j=1}^{N} w_{ij}^t y_{jt}$ is endogenous; thus, estimating the proposed model with the ML methods of non-spatial stochastic frontier models will generate a bias, unless $\rho = 0$. Taking into account the endogeneity of the spatial lag term, we present the ML methods to estimate the SARSF model.

Following Battese and Coelli [22], we re-parameterize as follows:

$$\sigma^2 := \sigma_v^2 + \sigma_u^2, \gamma := \frac{\sigma_u^2}{\sigma_v^2 + \sigma_u^2} \tag{3}$$

Then, the log-likelihood function in Eq. (1) is as follows:

$$\begin{aligned} LL(\boldsymbol{\beta}, \boldsymbol{\delta}, \gamma, \rho, \sigma^2; \boldsymbol{y}) = \ln|\boldsymbol{I}_{NT} - \rho \boldsymbol{W}| - \frac{1}{2}\left(\sum_{i=1}^{N} T_i\right)\left[\ln\sigma^2 + \ln 2\pi\right] \\ - \frac{1}{2}\sum_{i=1}^{N}\sum_{t=1}^{T_i}\left(\frac{\boldsymbol{z}_{it}'\boldsymbol{\delta} + y_{it} - \boldsymbol{x}_{it}'\boldsymbol{\beta} - \rho\sum_{j=1}^{N} w_{ij}^t y_{jt}}{\sigma}\right)^2 \\ - \sum_{i=1}^{N}\sum_{t=1}^{T_i}\left[\ln\Phi(d_{it}) - \ln\Phi\left(d_{it}^*\right)\right] \end{aligned}$$

$$\mu_{it}^* := \boldsymbol{z}_{it}'\boldsymbol{\delta}(1-\gamma) - \left(y_{it} - \boldsymbol{x}_{it}'\boldsymbol{\beta} - \rho\sum\nolimits_{j=1}^{N} w_{ij}^t y_{jt}\right)\gamma$$

$$\sigma^* := \sigma\sqrt{(1-\gamma)\gamma}, d_{it}^* := \frac{\mu_{it}^*}{\sigma^*}, d_{it} := \frac{\boldsymbol{z}_{it}'\boldsymbol{\delta}}{\sigma\sqrt{\gamma}}$$

Here, $\ln|\boldsymbol{I}_{NT} - \rho\boldsymbol{W}|$ comes from the Jacobian that accompanies the variable transformation from $\varepsilon_{it} := v_{it} - u_{it}$ to y_{it}, considering the endogeneity of the spatial lag term. We maximize the log-likelihood function numerically with this first-order condition satisfied (The first order conditions are Eqs. (13)–(17) in [1]).

As proposed by Battese and Coelli [31], the TE score TE_{it} is measured by the expectation of $\exp(-u_{it})$ conditional on $\varepsilon_{it} = v_{it} - u_{it}$.

$$TE_{it} := \mathrm{E}(\exp(-u_{it})|\varepsilon_{it}) = \exp\left[-\mu_{it}^* + \frac{1}{2}\sigma^{*2}\right] \cdot \left\{\frac{\Phi\left(\frac{\mu_{it}^*}{\sigma^*} - \sigma^*\right)}{\Phi\left(\frac{\mu_{it}^*}{\sigma^*}\right)}\right\} \tag{4}$$

The estimates of TE_{it} are obtained by evaluating Eq. (4) with the ML estimates.

4 Application to the Japanese Manufacturing Industry

In this section, using balanced regional panel data of all 47 prefectures in Japan during the 13 years from 2002 to 2014, we estimate the regional aggregate production function of the manufacturing industry. Each prefecture is one of the important policy makers on its industry. Some studies have applied spatial stochastic frontier models, especially to European data or country-level data (Fusco and Vidoli [32] in 2013; Vidoli et al. [33] in 2016 (Italy); Glass et al. [5] in 2016 (41 European countries); Ramajo and Hewings [13] in 2018 (9 Western European countries); Han et al. [34] in 2016 (21 OECD countries)). However, we do not know any study to apply a spatial stochastic frontier model to Japanese regional data.

The significance of this application is as follows. First, we clarify the features of our model and confirm the usefulness of the proposed model for empirical studies. Our model nests many existing spatial and non-spatial stochastic frontier models. By comparing our model and these existing models using real-world data, we check whether the model difference such as the consideration of the spatial interdependence, affects their estimates or efficiency scores. Second, we verify whether it is necessary to consider the spatial interdependence on the estimation of prefectural production function. There are many studies of estimating prefectural production functions. The existence of the example where the spatial interdependence is present means that spatial interdependence needs to be generally considered when estimating prefectural production functions. In this case, it is inappropriate to use models that do not consider the spatial interdependence unless we show the non-presence of spatial interdependence in the dataset by statistical test using a model considering spatial interdependence. Third, this study presents important policy implications for the manufacturing industry. This is analyzed in relation to the spatial interdependence, which can arise as a result of spatial externalities such as knowledge spillover and input–output networks. Each prefecture is one of the important policy makers about its own industrial policy, whereas the Japanese government is trying to form industrial clusters across a wide range of prefectures. The government expects positive spatial externalities across prefectures. Thus, it is important to the industrial policy to verify whether or not manufacturing industry has an interdependency relationship straddling prefectures.

We adopt the Cobb–Douglas production function, which is one of the most standard production functions. The estimation equation is as follows:

$$\ln y_{it} = \alpha + \beta_l \ln L_{it} + \beta_k \ln K_{it} + \beta_t t + \beta_{t^2} t^2 + \rho \sum_{j=1}^{N} w_{ij} y_{jt} + v_{it} - u_{it},$$
$$i = 1, 2, \ldots, 47, t = 0, 1, 2, \ldots, 12. \tag{5}$$

$$\mu_{it} = \delta_0 + \delta_1 DPOP_{it} + \delta_2 DPOP_{it}^2 + \delta_3 LARGE_{it} + \delta_4 WHOURS_{it} + \delta_5 WHOURS_{it}^2 \tag{6}$$

In the above equations, y_{it}, L_{it}, and K_{it} denote output, labor input, and capital input, respectively. β_l and β_k are the unknown estimated parameters about labor and capital, respectively. As with Glass et al. [22], assuming a Hicks-neutral technical change, we add a linear time trend variable t and its square (t is 0, with 2002 as the benchmark year, and increases by 1 for each year) in Eq. (5).

The measure of output is value-added (mn yen) in manufacturing establishments with 30 or more employees; the data is taken from the census of manufacture by the Ministry of Economy, Trade and Industry. Labor input is the number of workers multiplied by the working hours per capita. The number of employees in manufacturing establishments with 30 or more employees is taken from the industrial statistical survey. The working hours are the total average monthly hours worked per capita of regular employees in manufacturing establishments with 30 or more employees, and this data is taken from the monthly labor survey (regional survey) by the Ministry of Health, Labor and Welfare. Capital input is the "value of the tangible fixed assets, other than land," (mn yen) in manufacturing establishments with 30 or more employees from; this data is taken from the census of manufacture.[1]

Concerning the spillover, it is natural to think that the spatial interdependence, such as externalities, reduces with geographical distance. In fact, in many urban economic studies, the property that knowledge spillover decays with distance is thought to lead to the formation of cities, that is, agglomeration [35]. Thus, in this study, the spatial weight matrix is defined as row-normalized inverse distances between the prefectural offices (km). As mentioned in the model section, the row-normalization eliminates the influence of the distance unit.

The variables that represent the determinants of technical inefficiency include population density $DPOP_{it}$, the ratio of large establishments $LARGE_{it}$, and the per capita working hours $WHOURS_{it}$. $DPOP_{it}$ is obtained by dividing the intercensal adjusted population aged 15 to 64 years, as on October 1 of each year, by the inhabitable land of each prefecture (ha); this data is taken from the population estimates of the Ministry of Internal Affairs and Communications. This can be regarded as a proxy variable of agglomeration. An agglomeration economy is usually considered to have a positive influence on efficiency. However, as the agglomeration progresses, the effect of agglomeration may become negative because of a dominating congestion effect. Taking this into account, we also add the squared term of population density. $LARGE_{it}$ is the relative number of manufacturing establishments with 300 or more employees to those with 30 or more employees. Based on the estimates by aggregated data, it is impossible to distinguish whether the increasing returns to scale of the regional production function is because of the agglomeration economy or the increasing returns to scale at the establishment level. In this regard, $LARGE_{it}$ is expected to capture the economies of scale at the establishment level. The data required to compute $LARGE_{it}$ are taken from the census of manufacture. In Eq. (6), the squared term of $WHOURS_{it}$ is also an explanatory variable. This is because of the existence of optimal working hours. In Japan, the long working hours is a social problem from the viewpoint of health, work-life balance, and labor productivity. Therefore, some policies are underway to regulate working hours in order to

[1]The index variables of monetary value are not nominal, but real. For the manufacturing industry, we apply a chain-linked deflator to the value-added; for the private enterprise equipment, this deflator is also applied to the value of tangible fixed assets. These deflators are taken from "National Accounts for 2015" (2008 SNA, benchmark year = 2011).

improve the welfare of workers. However, from the perspective of labor productivity, too short working time per worker may be costly for managers to coordinate the production plan. We will measure the optimal working hours per worker in the next section.

Tables 1 and 2 show the descriptive statistics and correlation coefficient matrix of those variables in our dataset, respectively. There is a correlation between the explanatory and determinant variables of inefficiency. If the inefficiency determinants that correlate with explanatory variables are omitted, then the parameter estimate in the PPF will be biased.

The proposed SARSF model for panel data that incorporates a model of technical inefficiency (hereinafter referred to as SSFTE), nests many existing spatial and non-spatial stochastic frontier models. Therefore, in addition to the proposed model, we will estimate several models with constraints on parameters. First, the model with $\rho = \gamma = 0$ and $\boldsymbol{\delta} = 0$ is a linear regression model. Second, the model with $\gamma = 0$ and $\boldsymbol{\delta} = 0$ is an SAR regression. Third, the model with $\rho = 0$ and $\boldsymbol{\delta} = 0$ is a non-spatial stochastic frontier model with a half-normal distribution proposed by Aigner et al. [2] (hereinafter, ALS). Fourth, the model with $\boldsymbol{\delta} = 0$ is a spatial stochastic frontier model with an SAR structure and a half-normal distribution proposed by Glass et al.

Table 1 Summary statistics

	Max	Min	Mean	Mode	Std. dev
ln$y$	16.28	11.00	13.80	11.00	0.99
ln$L$	18.58	14.49	16.49	14.49	0.82
ln$K$	15.57	11.37	13.50	11.37	0.88
DPOP	10.93	0.28	1.84	1.65	2.12
LARGE	0.1131	0.0000	0.0664	0.0625	0.0189
WHOURS	175.90	151.10	165.71	166.70	4.43

Note y = output; L = labor input; K = capital input; *DPOP* = population density; *LARGE* = ratio of large establishments; *WHOURS* = per capita working hours.

Table 2 Correlation coefficient matrix

	ln$y$	ln$L$	ln$K$	*DPOP*	*LARGE*	*WHOURS*
ln$y$	1.000	0.514	0.528	0.234	0.276	0.028
ln$L$		1.000	0.384	0.153	−0.143	0.197
ln$K$			1.000	−0.234	0.095	−0.228
DPOP				1.000	−0.226	−0.356
LARGE					1.000	0.025
WHOURS						1.000

Note y = output; L = labor input; K = capital input; *DPOP* = population density; *LARGE* = ratio of large establishments; *WHOURS* = per capita working hours

[5] (hereinafter, GKS). Fifth, the model with $\rho = 0$ is a non-spatial stochastic frontier model that incorporates the model of technical inefficiency proposed by Battese and Coelli [22] (hereinafter, BC95).

5 Estimation Results

Table 3 shows the estimation results. In the models with the spatial lag, the coefficient ρ of the spatial lag is statistically significant at the 1% significance level, with a positive sign. This indicates that the production activities of the Japanese manufacturing industry are spatially dependent. The production activities have mutually positive effects. This lends support to the industrial cluster policy of the Japanese government. The magnitude of the coefficient varies depending on the models. The coefficient in SSFTE is smaller than that in the other models ($\rho = 0.3129$ in SAR and $\rho = 0.3329$ in GKS; however, $\rho = 0.2115$ in SSFTE). In models that do not consider determinants of technical inefficiency, ρ is considered to be overestimated such that ρ absorbs some of the heteroskedasticity of the technical inefficiency. This indicates the necessity to incorporate the determinants of technical inefficiency into the spatial models.

Table 4 shows the labor elasticity of production, the capital elasticity of production, the degree of the returns to scale, and the Hicks-neutral technical change rate, which are calculated from the estimation results. In the model with the spatial lag term, those values vary over observations, and hence we have displayed their maximum, minimum, and average values. The average values are equivalent to the summary statistics of the direct effect in [20]. The degree of the returns to scale is the sum of the labor elasticity of production and the capital elasticity of production. The degree of returns to scale greater (less) than 1 indicates increasing (decreasing) returns to scale technology.

The labor coefficient and the labor elasticity in the models with a spatial lag are lower than those in the models without a spatial lag. This suggests an overestimation of the labor elasticity value in the model without a spatial lag, as labor input correlates with the spatial spillover effects, including externality. The degrees of returns to scale indicate the economy of scale in not only linear regression and SAR but also in the stochastic frontier models without a model of technical efficiency. For example, the estimates by ALS and GKS are 1.08 and 1.15, respectively, indicating increasing returns to scale. However, BC95 and SSFTE show almost constant returns to scale technology, as their estimates of the degree of the returns to scale are 1.005 and 1.01, respectively. This suggests that the coefficients of input quantities in models that ignore the determinants of technical inefficiency are overestimated because of the correlation between the determinants of technical inefficiency and the input amount, especially the capital input.

As Table 3 shows, in all the models, the coefficient of the time trend in the PPF is statistically significant at the 5% significance level, and its sign is positive. The sign of the coefficient of the squared time trend is negative, but insignificant for all the

Table 3 Estimation results

	OLS		SAR		ALS		GKS		BC95		SSFTE	
	Coef	z-stat	Coef	z-stat	Coef	z-stat	Coef	z-stat	Coef	z-stat	Coef	z-stat
α	−3.6458***	(−18.95)	−7.0903***	(−16.89)	−3.1177***	(−13.66)	−6.5592***	(−16.15)	−1.2218***	(−5.24)	−3.9661***	(−7.95)
β_l	0.5987***	(18.31)	0.5100***	(15.89)	0.5483***	(16.10)	0.4180***	(13.44)	0.5396***	(19.05)	0.4654***	(14.88)
β_k	0.5501***	(17.95)	0.5943***	(20.67)	0.5854***	(19.17)	0.6613***	(25.63)	0.4651***	(15.91)	0.5382***	(16.25)
β_t	0.0327***	(3.75)	0.0175**	(2.17)	0.0322***	(4.23)	0.0151**	(1.99)	0.0369***	(4.87)	0.0239***	(3.28)
β_{t^2}	−0.0010	(−1.39)	−0.0001	(−0.22)	−0.0010	(−1.64)	0.0000	(0.00)	−0.0014**	(−2.28)	−0.0006	(−1.04)
δ_0									49.1749***	(4.53)	45.0132***	(3.47)
δ_1									−0.1961***	(−6.18)	−0.1746***	(−4.51)
δ_2									0.0079*	(1.83)	0.0041	(0.57)
δ_3									−7.6464***	(−9.35)	−8.1727***	(−6.87)
δ_4									−0.5770***	(−4.37)	−0.5280***	(−3.35)
δ_5									0.0017***	(4.31)	0.0016***	(3.30)
ρ			0.3129***	(9.16)			0.3329***	(10.04)			0.2115***	(6.20)
γ					0.6119***	(6.66)	0.7463***	(11.33)	0.7692***	(14.53)	0.7334***	(14.51)
σ^2	0.0468***	(17.81)	0.0415***	(17.65)	0.0765***	(8.75)	0.0785***	(9.13)	0.0496***	(10.97)	0.0529***	(7.05)
LL	68.1950		103.9273		72.3177		114.3961		163.8507		182.7421	
AIC	−124.3900		−193.8547		−130.6353		−212.7922		−301.7013		−337.4843	

Note OLS: Linear Regression; SAR: spatial autoregressive model; ALS: non-spatial stochastic frontier model with half-normal distribution proposed by Aigner et al. [2], GKS: spatial stochastic frontier model with a SAR structure and a half-normal distribution proposed by Glass et al. [5], BC95: non-spatial stochastic frontier model that incorporates a model of technical inefficiency proposed by Battese and Coelli [22] SSFTE: our proposed model; Standard error is calculated using inverse the numerically estimated negative Hessian of the log likelihood evaluated at the maximum likelihood estimators; *, **, and *** denote statistical significance at the 10%, 5%, and 1% levels, respectively

Table 4 Marginal effects and technological progress rate

	OLS	SAR	ALS	GKS	BC95	SSFTE
Frontier			x	x	x	x
Spatial lag		x		x		x
Model of TE					x	x
Labor elasticity of production						
Max.	0.5987	0.5171	0.5483	0.4247	0.5396	0.4682
Min.	0.5987	0.5105	0.5483	0.4184	0.5396	0.4655
Mean	0.5987	0.5129	0.5483	0.4207	0.5396	0.4665
Capital elasticity of production						
Max.	0.5501	0.6025	0.5854	0.6719	0.4651	0.5415
Min.	0.5501	0.5948	0.5854	0.662	0.4651	0.5384
Mean	0.5501	0.5976	0.5854	0.6656	0.4651	0.5395
Degrees of returns to scale						
Max.	1.149	1.120	1.134	1.097	1.005	1.010
Min.	1.149	1.105	1.134	1.080	1.005	1.004
Mean	1.149	1.111	1.134	1.086	1.005	1.006
Hicks-neutral technical change rate						
Max.	0.0209	0.0167	0.0204	0.0155	0.0203	0.0169
Min.	0.0209	0.0164	0.0204	0.0152	0.0203	0.0168
Mean	0.0209	0.0165	0.0204	0.0153	0.0203	0.0169

Note Model of TE: model of technical inefficiency; OLS: Linear Regression; SAR: spatial autoregressive model; ALS: non-spatial stochastic frontier model with a half-normal distribution proposed by Aigner et al. [2]; GKS: spatial stochastic frontier model with a SAR structure and a half-normal distribution proposed by Glass et al. [5]; BC95: non-spatial stochastic frontier model that incorporates a model of technical inefficiency proposed by Battese and Coelli [22]; SSFTE: our proposed model; Hicks-neutral technical change rate is the mean annual technical progress rate calculated based on the estimated time trends

models, except BC95. The results indicate that the PPF shifts upward by a technical change during the period of the analysis; in this case, the rate of shift of the PPF is constant or decreasing. The Hicks-neutral technical change rate in Table 4 shows that the change rate in the models considering the spatial interdepende nce (SAR, GKS and SSFTE) is lower than that in the models that do not consider spatial interdependence. The coefficient of population density is significantly negative at the 1% significance level in both BC95 and SSFTE. The sign of the coefficient of the square term of the population density is also positive in both BC95 and SSFTE. In BC95, it is statistically significant at the 10% significance level, while it is not significant in the case of SSFTE. Eventually, it is implied that the increase in population density raises technical efficiency within the dataset. In BC95 and SSFTE, the coefficients of working hours and working hours' squares are statistically significant at the 1% significance level. The per capita working hours to maximize technical efficiency

Table 5 LR test results

Model	Null hypothesis	Number of constraints	Test statistic	1% rejection statistic	Decision
OLS	$\rho = \gamma = 0$, $\boldsymbol{\delta} = 0$	8	229.09	26.12	Reject
SAR	$\gamma = 0, \boldsymbol{\delta} = 0$	7	157.63	24.32	Reject
ALS	$\rho = 0, \boldsymbol{\delta} = 0$	7	220.85	24.32	Reject
GKS	$\boldsymbol{\delta} = 0$	6	136.69	22.46	Reject
BC95	$\rho = 0$	1	37.78	10.83	Reject

Note ALS: non-spatial stochastic frontier model with half-normal distribution proposed by Aigner et al. [2]; GKS: spatial stochastic frontier model with a SAR structure and a half-normal distribution proposed by Glass et al. [5]; BC95: non-spatial stochastic frontier model that incorporates a model of technical inefficiency proposed by Battese and Coelli [22]; SSFTE: our proposed model

were 167.1 h and 167.2 h for BC95 and SSFTE, respectively. Thus, this result is robust in the specification of the model. The coefficients of the large-scale business establishment ratio are significantly negative at the 1% significance level in BC95 and SSFTE. This suggests the existence of the economies of scale at the establishment level.

Table 5 shows the results of the likelihood ratio (LR) test on several models nested by the proposed model.[2] The null hypothesis of no spatial lag (i.e., $\mathrm{H_o} : \rho = 0$) is rejected at the 1% significance level. The model with a spatial lag is empirically supported. The null hypothesis of no determinants of technical inefficiency (i.e., $\mathrm{H_o} : \boldsymbol{\delta} = 0$) and the null hypothesis of no spatial lag and no determinants of technical efficiency (i.e., $\mathrm{H_o} : \rho = 0 \text{and} \boldsymbol{\delta} = 0$) are both rejected at the 1% significance level. The modeling determinants of technical inefficiency are statistically supported. The null hypothesis of no technical inefficiency (i.e., $\mathrm{H_o} : \gamma = 0 \text{and} \boldsymbol{\delta} = 0$) is rejected at the 1% significance level. This supports the composed error structure peculiar to the stochastic frontier model. The null hypothesis of no spatial lag and no determinants of technical efficiency and no technical inefficiency (i.e., $\mathrm{H_o} : \rho = \gamma = 0 \text{and} \boldsymbol{\delta} = 0$) is decisively rejected at the 1% significance level. As a result of the LR test, all the existing nested models are rejected, indicating that SSFTE is preferable.

Next, we compare the TE score in the several models. Although there are several definitions of the TE score, in order to compare the effect of the estimation models, we unify them by defining them, as in Eq. (4). Figure 1 shows the TE scores' histogram using 611 observations. The average of the TE scores in SSFTE, BC95, GKS, and ALS are 0.8047, 0.7530, 0.8494, and 0.8348, respectively. In the model where the determinants of technical inefficiency are considered, the distribution of TE is dispersed. By considering the spatial interdependence, that is, removing the constraint of $\rho = 0$, it is found that the TE score tends to approach 1.

[2]The LR test statistic is defined as $LR_\lambda = -2\{LL[H_1] - LL[H_0]\}$, where $LL[H_1]$ and $LL[H_0]$ are the log-likelihood functions under H_1 and H_0, respectively. This test statistic asymptotically follows the chi-square distribution with degrees of freedom equal to the number of constraints.

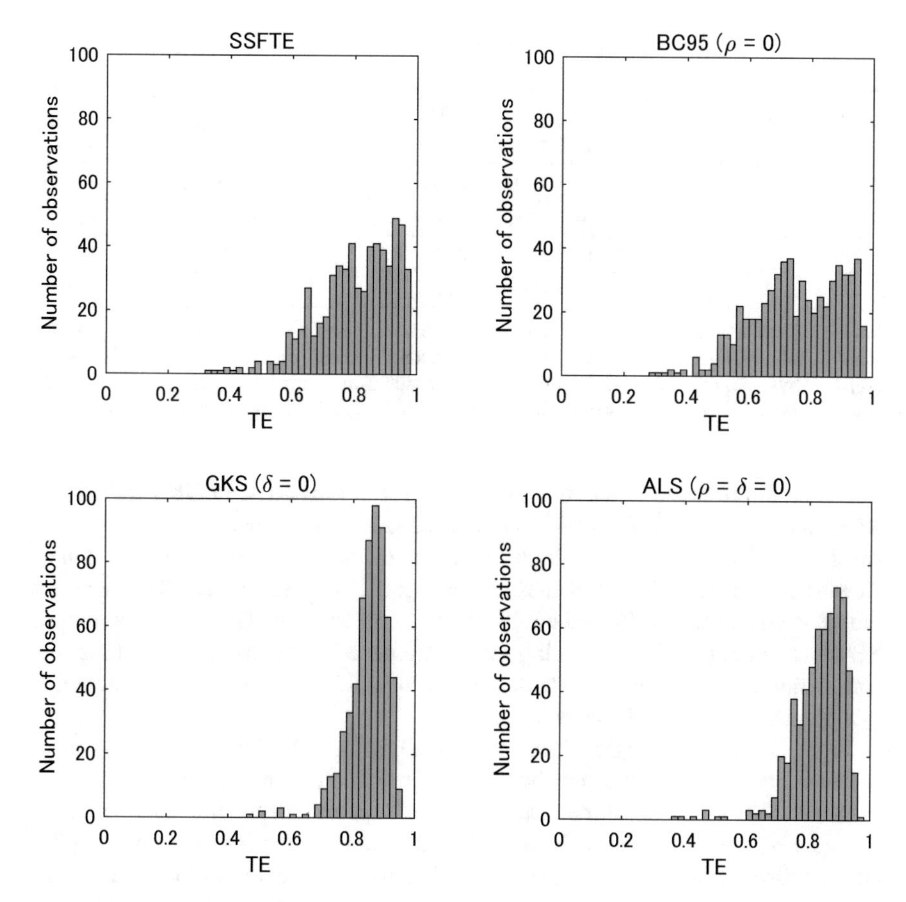

Fig. 1 Distribution of TE scores

Table 6 shows the period mean of Spearman's rank correlation coefficient (SRCC) matrix and the period mean of the maximum rank difference ratio (MRDR). Let the TE score ranking of the ith producer in the period t, in the model K, be R_{it}^{K}, then the MRDR of models A and B will be as follows:

$$MRDR_{ABt} := \frac{\max_{i}\left|R_{it}^{A} - R_{it}^{B}\right|}{N} \tag{7}$$

As expected, there are positive correlations between the TE score ranking in all models. However, the TE score ranking changes significantly between models using variables that explain the determinants of technical inefficiency (SSFTE and BC 95) and models that do not use those variables (GKS and ALS). Given that there is statistical significance between the variables describing the determinants of technical inefficiency, it is important to consider the determinants of technical inefficiency when conducting an estimation. The presence or absence of the spatial lag does not

Table 6 Spearman's rank correlation coefficient (SRCC) and maximum rank difference ratio (MRDR)

Mean SRCC					Mean MRDR				
	SSFTE	BC95	GKS	ALS		SSFTE	BC95	GKS	ALS
SSFTE	1.000	0.980	0.814	0.798	SSFTE	0.000	0.177	0.512	0.604
BC95		1.000	0.789	0.720	BC95		0.000	0.506	0.622
GKS			1.000	0.936	GKS			0.000	0.383
ALS				1.000	ALS				0.000

Note ALS: non-spatial stochastic frontier model with half-normal distribution proposed by Aigner et al. [2]; GKS: spatial stochastic frontier model with a SAR structure and a half-normal distribution proposed by Glass et al. [5]; BC95: non-spatial stochastic frontier model that incorporates a model of technical inefficiency proposed by Battese and Coelli [22]; SSFTE: our proposed model.

lead to so dramatic change in the rank order. The mean SRCC of GKS and ALS is 0.936, and the mean SRCC of SSFTE and BC95 is 0.980. Since SSFTE and BC 95 specify the determinants of technical inefficiency, the TE score ranking is similar. However, the mean MRDR of these models is 0.177, which means that there is a difference of up to 17.7% in the rank order, on an average. This indicates that the TE score ranking varies depending on the presence of the spatial lag. Taking into consideration the results and statistical test results, it is clear that the introduction of spatial lags plays a significant role.

Figure 2 shows the regional mean of the TE scores. As the overall mean of the scores varies by model, we map them using six quantiles; this mapping is attributed to the fact that the shape of the distribution varies greatly depending on the models. Considering the discussion so far, the efficiency scores differ depending on the presence or absence of a spatial lag and the determinants of technical inefficiency. Let us consider the area around the Aichi Prefecture, where the automobile industry agglomerates and where the value-added is the largest. The ranking of the TE score of the prefectures around the Aichi Prefecture (e.g., the Shizuoka and Gifu prefectures) in SSFTE is lower than that in BC95. In these areas, it is considered that the efficiency scores decrease because of the introduction of the spatial lag, which makes the PPF shift upward around the Aichi prefecture.

6 Conclusions

We developed a spatial stochastic frontier model with the SAR term and the feature of Battese and Coelli's model [22], which simultaneously estimates the determinants of technical inefficiency. Subsequently, we conducted an empirical analysis using data from the Japanese manufacturing industry. Statistical tests have supported the proposed model. We found that the production activities of the Japanese manufacturing industry are spatially interdependent and produce mutually positive effects.

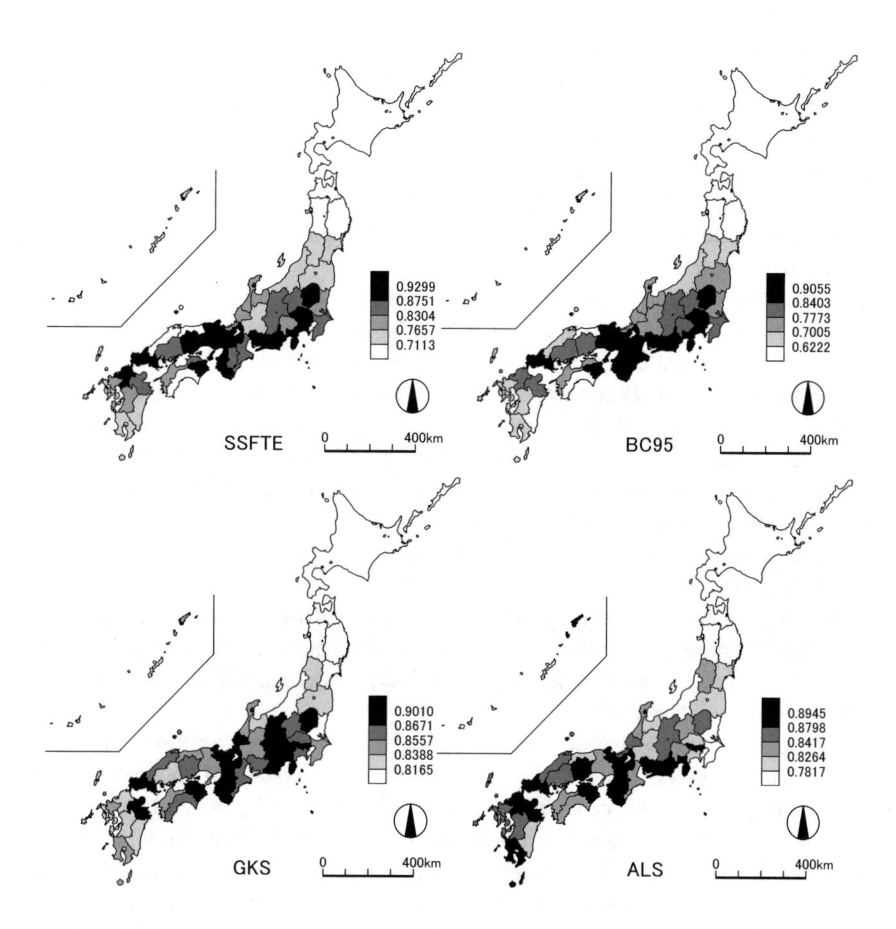

Fig. 2 Regional mean of TE scores (six quantiles)

This lends support to the industrial cluster policy of the Japanese government. Our findings suggest that, in the existing spatial and non-spatial models, the estimates (e.g., labor and capital elasticities and spatial interdependence) in the existing spatial and non-spatial models were biased because of a lack of consideration to the determinants of technical inefficiency and the spatial lag. This bias also affected the TE score and its ranking.

In particular, it is a significant conclusion that, in models without the determinants of technical inefficiency, the scale parameter of spatial interdependence ρ is overestimated because ρ absorbs some of the heteroskedasticity of technical inefficiency. This implies an over-measurement of the spatial spillover, such as externalities, and it can lead to an erroneous policy judgment. Our model can measure the spatial spillover while controlling for the heteroskedasticity of technical inefficiency; in this respect, our model is superior to the aforementioned models.

Using the proposed model, we can statistically ascertain the presence of spatial interdependence and the need to consider the determinants of technical inefficiency. Therefore, if the test supports spatial independence, then studies can use the existing non-spatial stochastic frontier models such as BC95. If the test supports that the determinants of technical inefficiency are not required, then studies can use the existing spatial stochastic frontier model such as GKS. There is no positive reason to first use models that do not consider both of the spatial interdependence and the determinants of technical inefficiency.

Our model has some extensibility. We can easily introduce a spatial lag of explanatory variables into our model. In the field of spatial econometrics, the model that adds both the spatial lags of the explained and explanatory variables is called the spatial Durbin model. Therefore, this extended model can be called a spatial Durbin stochastic frontier model that incorporates a model of technical inefficiency. Since these added variables are all exogenous, we can estimate this model directly using our estimation method. We can also potentially introduce the SEM structure into the error term in our model. The SEM structure can address the spatial interdependence in the error term. By adding the additional weight matrix, we can extend our model to higher order spatial econometric models [36, 37].

Recently, there are many studies on how to deal with endogenous explanatory variables in stochastic frontier models [38, 39]. However, it is difficult to apply the methods to spatial interdependence models, including our model. This remains as a future subject of research.

In this study, we proposed a useful spatial stochastic frontier model; however, several challenges and applicability remain from the viewpoint of application. First, we confirmed the existence of the spatial interdependence by using the prefectural data because each prefecture is a policymaker about manufacturing industry. Hence, from a policy standpoint, it is important to exhibit the spatial interdependence across prefectures. Our proposed model also allows various other analyses on spillover. For example, if an analyst is interested in interdependence relationships between firms, conducting a firm-level analysis using the proposed model may results in new findings. Second, the proposed model can reduce the omitted-variable bias by introducing appropriate determinants of technical inefficiency. We expect further research on the appropriate determinants of technical inefficiency. Finally, we used the geographical distance for the spatial weight matrix, whereas by creating a spatial weight matrix based on the economic distance calculated using the input–output tables, it is possible to analyze in consideration of technological proximity [40]. As described above, there is much room for empirical studies. The proposed model is expected to be applied to empirical analysis in many fields, including regional science and productivity analysis.

Finally, we present an example of a potential RWDC about this study (Fig. 3). Consider an example of the Japanese manufacturing industry. The data required for productivity analysis of the Japanese manufacturing industry are continuously obtained through surveys, such as the census of manufacture and the economic census. Analysis can be performed using these data and our proposed model. In particular, the current industrial policy will be evaluated, and also it is expected that,

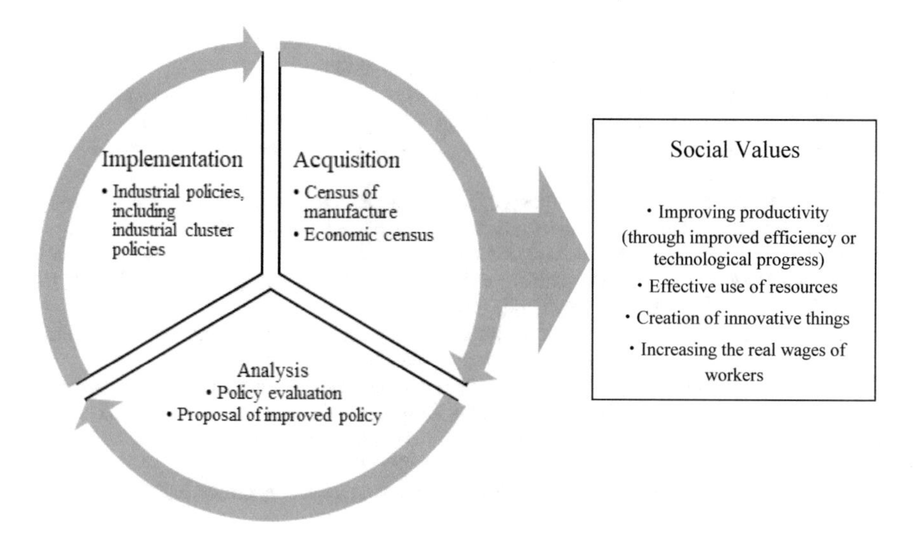

Fig. 3 Potential Real-World Data Circulation about our Study

by devising the spatial weight matrix, it will be possible to analyze what kind of industrial agglomeration form is desirable for improving productivity. Based on the results of the analysis, we will implement industrial policies, including industrial cluster policies. It is expected that the policy proposals and implementations can be made effective by acquiring and analyzing data after the "intervention" of the policy implementation.

References

1. Tsukamoto T (2019) A spatial autoregressive stochastic frontier model for panel data incorporating a model of technical inefficiency. Japan World Econ 50:66–77
2. Aigner D, Lovell CK, Schmidt P (1977) Formulation and estimation of stochastic frontier production function models. J Econ 6(1):21–37
3. Meeusen W, van Den Broeck J (1977) Efficiency estimation from Cobb-Douglas production functions with composed error. Int Econ Rev 18(2):435–444
4. Elhorst JP (2014) Spatial econometrics from cross-sectional data to spatial panels. Springer
5. Glass AJ, Kenjegalieva K, Sickles RC (2016) A spatial autoregressive stochastic frontier model for panel data with asymmetric efficiency spillovers. J Econom 190(2):289–300
6. LeSage JP, Pace RK (2009) Introduction to spatial econometrics. Chapman and Hall/CRC, Boca Raton
7. Pace RK, Zhu S (2012) Separable spatial modeling of spillovers and disturbances. J Geogr Syst 14(1):75–90
8. Fingleton B, López-Bazo E (2006) Empirical growth models with spatial effects. Papers Region Sci 85(2):177–198
9. Elhorst JP (2010) Applied spatial econometrics: raising the bar. Spat Econ Anal 5(1):9–28
10. Druska V, Horrace WC (2004) Generalized moments estimation for spatial panel data: Indonesian rice farming. Am J Agr Econ 86(1):185–198

11. Kelejian HH, Prucha IR (1999) A generalized moments estimator for the autoregressive parameter in a spatial model. Int Econ Rev 40(2):509–533
12. Adetutu M, Glass AJ, Kenjegalieva K, Sickles RC (2015) The effects of efficiency and TFP growth on pollution in Europe: a multistage spatial analysis. J Prod Anal 43(3):307–326
13. Ramajo J, Hewings GJ (2018) Modelling regional productivity performance across Western Europe. Regional Stud 52(10):1372–1387
14. Battese GE, Coelli TJ (1992) Frontier production functions, technical efficiency and panel data: With application to paddy farmers in India. J Prod Anal 3(1–2):153–169
15. Gude A, Álvarez IC, Orea L (2017) Heterogeneous spillovers among Spanish provinces: a generalized spatial stochastic frontier model (No. 2017/03). University of Oviedo, Department of Economics, Oviedo Efficiency Group (OEG)
16. Kalirajan K (1981) An econometric analysis of yield variability in paddy production. Canadian J Agric Econ/Revue Canadienne D'Agroeconomie 29(3):283–294
17. Kumbhakar SC, Ghosh S, McGuckin JT (1991) A generalized production frontier approach for estimating determinants of inefficiency in US dairy farms. J Busin Econ Statist 9(3):279–286
18. Reifschneider D, Stevenson R (1991) Systematic departures from the frontier: a framework for the Analysis of firm inefficiency. Int Econ Rev 32(3):715–723
19. Caudill SB, Ford JM (1993) Biases in frontier estimation due to heteroscedasticity. Econ Lett 41(1):17–20
20. Huang CJ, Liu JT (1994) Estimation of a non-neutral stochastic frontier production function. J Prod Anal 5(2):171–180
21. Caudill SB, Ford JM, Gropper DM (1995) Frontier estimation and firm-specific inefficiency measures in the presence of heteroscedasticity. J Busin Econ Statist 13(1):105–111
22. Battese GE, Coelli TJ (1995) A model for technical inefficiency effects in a stochastic frontier production function for panel data. Empirical Econ 20(2):325–332
23. Kumbhakar SC, Lovell CK (2003) Stochastic frontier analysis. Cambridge University Press, Cambridge
24. Fries S, Taci A (2005) Cost efficiency of banks in transition: evidence from 289 banks in 15 post-communist countries. J Bank Finance 29(1):55–81
25. Srairi SA (2010) Cost and profit efficiency of conventional and Islamic banks in GCC countries. J Prod Anal 34(1):45–62
26. Saeed M, Izzeldin M (2016) Examining the relationship between default risk and efficiency in Islamic and conventional banks. J Econ Behav Organ 132:127–154
27. Pavlyuk D (2011) Application of the spatial stochastic frontier model for analysis of a regional tourism sector. Transport Telecommun 12(2):28–38
28. Arbia G (2014) A primer for spatial econometrics: with applications in R. Springer.
29. Kelejian HH, Tavlas GS, Hondroyiannis G (2006) A spatial modelling approach to contagion among emerging economies. Open Econ Rev 17(4–5):423–441
30. Stevenson RE (1980) Likelihood functions for generalized stochastic frontier estimation. J Econom 13(1):57–66
31. Battese GE, Coelli TJ (1988) Prediction of firm-level technical efficiencies with a generalized frontier production function and panel data. J Econom 38(3):387–399
32. Fusco E, Vidoli F (2013) Spatial stochastic frontier models: controlling spatial global and local heterogeneity. Int Rev Appl Econ 27(5):679–694
33. Vidoli F, Cardillo C, Fusco E, Canello J (2016) Spatial nonstationary in the stochastic frontier model: An application to the Italian wine industry. Regional Sci Urban Econom 61:153–164
34. Han, J., Ryu, D., and Sickles, R.C., (2016). Spillover effects of public capital stock using spatial frontier analyses: A first look at the data. In W.H. Green, L. Khalaf, R. Sickles, M. Veall, M-C. Voia (Eds.), *Productivity and Efficiency Analysis* (pp. 83–97). Springer Proceedings in Business and Economics. Cham: Springer.
35. Marshall A (1890) Principles of economics. Macmillan, London
36. Lacombe DJ (2004) Does econometric methodology matter? an analysis of public policy using spatial econometric techniques. Geogr Anal 36(2):105–118

37. Elhorst JP, Lacombe DJ, Piras G (2012) On model specification and parameter space definitions in higher order spatial econometric models. Regional Sci Urban Econom 42(1–2):211–220
38. Kutlu L (2010) Battese-Coelli estimator with endogenous regressors. Econ Lett 109(2):79–81
39. Amsler C, Prokhorov A, Schmidt P (2017) Endogenous environmental variables in stochastic frontier models. J Econom 199(2):131–140
40. Dietzenbacher E, Luna IR, Bosma NS (2005) Using average propagation lengths to identify production chains in the Andalusian economy. Estudios De Economía Aplicada 23(2):405–422
41. Tsukamoto T (2019) Endogenous inputs and environmental variables in Battese and Coelli's (1995) stochastic frontier model. Econom Bull 39(3):2228–2236

Printed in the United States
by Baker & Taylor Publisher Services